Bruce Hood is currently the Director of the Bristol Cognitive Development Centre in the Experimental Psychology Department at the University of Bristol. He has been a research fellow at Cambridge University and University College London, a visiting scientist at MIT and a faculty professor at Harvard.

SUPERSENSE

From Superstition to Religion –
the Brain Science of Belief

Bruce Hood

Constable • London

Constable & Robinson Ltd
3 The Lanchesters
162 Fulham Palace Road
London W6 9ER
www.constablerobinson.com

First published in the US by HarperOne, a division of HarperCollins Publishers, 2009
This edition first published in the UK by Constable,
an imprint of Constable & Robinson Ltd, 2009

A copy of the British Library Cataloguing in
Publication data is available from the British Library

ISBN: 978-1-84901-030-6

Printed and bound in the EU

1 3 5 7 9 10 8 6 4 2

CONTENTS

PROLOGUE:

WHY DO WE DEMOLISH EVIL HOUSES?

THE HOUSE AT 25 Cromwell Street, Gloucester, England, is no longer there. In October 1996, the city council ordered the removal of all physical traces of the Wests' home where young girls were raped, tortured and murdered by Fred and Rosemary during the 1970s. Fred had used his builder's skills to conceal the bodies at the three-storey family home. First he buried them under the basement floor but when he ran out of space, he turned to the garden. His own sixteen-year-old daughter, Heather, was entombed under the newly laid patio. During the investigation, there was a rumour that some of the paving stones had been stolen from the crime scene. Unscrupulous locals had salvaged the slabs and an unwitting resident was now the proud owner of a barbecue made from the stones used to hide the horrors at Cromwell Street.[1] Nick, a fifty-something landlord who owned other houses in the street, told me this rumour was a myth. He was there. The council had removed every last brick. These were crushed into dust and then scattered across a landfill site in unmarked locations.

In the brilliant sunshine of Holy Thursday, April 2007, I stood on the exact spot where many of the bodies had been buried. Nick helped me locate this. It's now a passageway between the remaining row of houses and a Seventh-Day Adventist Church. I did not know about this oddity of street planning and was shocked by the closeness of heaven to hell on earth. Could the congregation ever have imagined

1

what was going on next door as they prayed? Did this proximity to the church heighten the Wests' sense of depravity?

I watched for half an hour as Gloucester's youngsters used the convenient walkway to get to wherever they were going. Most were heading to the nearby park. The unseasonably hot April day had brought out loose summer clothing, carefree laughter, and a spring in the step of the youth. Very unusual for this grim, English city, well past its prime. As they sauntered past the overdressed psychology professor who seemed to be oddly preoccupied with a passageway, they were oblivious to the human suffering and atrocities committed at this spot thirty years earlier. And why not? It was simply an empty space.

Why do we demolish and remove houses associated with appalling murders? The same happened to the Oxford Apartments in Milwaukee, Wisconsin, where Jeffrey Dahmer lived, and the house where Ian Huntley murdered the two little girls in Soham, England. Dahmer's place is now a car park and 5 College Close has been laid to turf. Houses associated with notorious murders are difficult to resell. The Colorado home where the body of the child beauty star JonBenét Ramsey was found has been on and off the market, always selling below its true value. US realtors call these properties 'stigmatized homes' and they present a considerable marketing challenge. Disclosure laws vary from state to state. In Massachusetts, if you don't ask, they don't need to tell. In Oregon, vendors don't have to reveal anything. Hawaiian realtors are legally bound to reveal everything that might affect the value of a property, including ghosts.[2] In the United Kingdom, you have to declare whether you have fallen out with the neighbours in a dispute. But there is no legal requirement to tell prospective buyers about the murderous history of a house. Deception is common, since most people would prefer to see these places obliterated from existence and memory.

In 2000, Alan and Susan Sykes sat down to watch a Channel 5 TV documentary about Dr Samson Perera, the Leeds University scientist who murdered and dismembered his teenage daughter, 15 years earlier. As the programme unfolded, Alan and Susan were shocked to discover that their modest house in Wakefield, West Yorkshire, was the actual scene of the horrific act and that police had never fully recovered all of the 100 body parts that the poor girl was hacked into. Alan and Susan were distraught. They immediately moved out of the house of horror, selling it six months later at £8,000 less than they had originally paid for it. They would never have bought the house if they had known its gruesome past. When they tried to sue the previous owners, who knew of the murder, for not telling them, the judge was sympathetic but rejected their claim, as the vendors were not obliged to disclose this information. The question on the vendor's form asked, "Is there any information which you think the buyer may have the right to know?" Apparently, missing body parts secreted around the house were not considered significant information from a legal point of view.[3]

Could you live in the house in Wakefield? Even if there were no missing body parts secreted around the building, just the thought of something horrible taking place is enough to keep most away. Are you someone who would cross the street to avoid standing on the spot where evil took place or would you relish the thrill? Why do we feel the need to replace something with nothing?

A physical building is a powerful reminder that can trigger painful memories and emotions. Maybe I was no better than the trail of ghoulish sightseers to Cromwell Street that Nick had witnessed over the years. If there is nothing to look at, then shouldn't this keep the weirdos away? At least removing the visible reminder makes it easier for a community to heal and forget. But demolishing a building, crushing the rubble into dust and taking it away to secret locations with demolishers under oath not to reveal the final whereabouts seems a bit excessive.[4]

What would motivate a souvenir hunter to want to own a brick or some physical thing associated with a murderer? The same goes for objects such as Nazi memorabilia. The world's largest auction website, eBay, has banned the sale of these items and anything that glorifies hatred, violence, or intolerance. But what attracts people to them in the first place? Maybe it's the excitement of being subversive. Any parent with a rebellious teenager knows that the macabre is a source of fascination for these fledgling adults. Part of growing up is the need

FIG. 1: The passageway at 25 Cromwell Street where the Wests buried many of their victims. AUTHOR'S COLLECTION.

to express individuality through statements of rebellion. By their nature, taboo topics intrigue the young who want to be outrageous in an effort to shock.

What about collectors of less insidious memorabilia? Mature adults will pay good money for personal items that once belonged to famous people. Some are just common objects, but collectors covet them because of their connection with celebrities. Why else would someone bid on eBay for a fragment of bed linen that was once slept on by Elvis Presley? Were John Lennon's handwritten lyrics to 'Give Peace a Chance' really worth £420,000 at a Christie's sale in 2008?[5] Why pay £1,000 for a swatch of cloth taken from Princess Diana's wedding dress?[6] The charity website www.clothesoffourback.com, started by *Malcolm in the Middle* mom Jane Kaczmarek and *West Wing* actor Bradley Whitford, auctions clothes worn by celebrities for the benefit of children's charities. Many of these items were worn at award ceremonies such as the Oscars or Emmys. These events take place under the glare of the media spotlight, and even the stars most likely to win must sweat a little in anticipation as that envelope is opened. However, their tainted tuxedoes and grubby gowns are highly desirable to the general public. The charity used to offer a dry-cleaning option to successful auction bidders but eventually dropped the service as no one wanted the clothing washed. Maybe the bidders thought they could get the clothes cleaned more cheaply themselves. This seems unlikely, however, if the money was for charity. Why not clean secondhand clothes? After all, we usually wash our own clothes when they get sweaty. I think the real answer could be that collectors did not necessarily want to wear them. They wanted to own something intimate and personal to their idols and the more connection, the better. It's a fetish in the original use of the word: a belief that an object has supernatural powers.[7]

Memorabilia collectors and those with object fetishism are behaving in a very peculiar way. They are attributing to physical objects invisible properties that make them unique and irreplaceable. This kind of thinking is misguided. For one thing, significant objects can be faked. That brick, that tuxedo or that bed linen may be a forgery. In the Middle Ages, there was a roaring trade in Christian relics to cater to the legions of pilgrims traipsing across Europe from one holy shrine to the next. Relics could be any objects connected intimately with religious celebrities. Bones belonging to saints and martyrs were particularly popular as were any items connected with Jesus. Bits of the cross or shreds of the shroud were easy to fake and trade was brisk. If all the fragments of the crucifixion cross were put back together, there would probably be enough to build an ark. The professional sceptic James Randi recounts how, as a boy growing up in Montreal, he visited the St Joseph's Oratory shrine where the beatified monk Brother André Bessette once lived. Brother André was known as the miracle-worker of Mount Royal. Pilgrims would flock to the shrine seeking supernatural healing for all manner of ills and could reach in to touch the jar containing the preserved heart of the monk housed behind a metal grill in an ornate cabinet. Randi recalls how his father and godfather were asked one day by the proprietors of St Joseph's Oratory to cut up a roll of black gabardine fabric purchased from a local store into small squares. These were then sold in the gift shop as pieces of Brother Andre's actual robes worn on his deathbed. Maybe this early experience had a profound influence on Randi becoming a sceptic.[8]

Even if an object is inauthentic, many people treat such items as if they possess some property inherited from the previous owner. A property that defies scientific measure. Some believe such objects harbour an inner reality or essence that makes them unique and irreplaceable. Yes, these houses and objects have a history and yes, they may remind us of events and people, but many believe or more importantly *act* as if these essences are physical, tangible realities. Something to touch or something to avoid. But, of course, they are not. Sweat and blood may have DNA but not bricks and mortar from a house. Rather there is something else that we sense in these objects. Something supernatural.

SUPERSENSE

This book is about the origins of supernatural beliefs, why they are so common, and why they may be so difficult to get rid of. I believe the answer to each of these questions can be found in human nature and, in particular, the developing mind of the child. I am proposing that humans have a natural, intuitive way of reasoning that leads them to supernatural beliefs. Almost like a common sense, but one that is based on supernatural thinking. So let's call it our 'supersense'.

Throughout this book, I hope to convince you that we are naturally inclined towards supernatural beliefs. Many highly educated and intelligent individuals experience a powerful sense that there are patterns, forces, energies, and entities operating in the world that are denied by science because they go beyond the boundaries of natural phenomena we currently understand. More importantly, such experiences are not substantiated by a body of reliable evidence, which is why they are *super*natural and unscientific. The inclination or sense that they may be real is our supersense.

Why are humans so willing to entertain the possibility of the supernatural? As we will see, most people believe because they think they have experienced supernatural events personally, or they have heard reliable testimony about the supernatural from those they trust. I would argue that we interpret our experiences and other peoples' reports within a supernatural framework because that framework is one that is intuitively plausible. It resonates with the way we think the world operates with all manner of hidden structures and mechanisms. If this is true, we have to ask where does this supersense come from?

Some argue that the most obvious origins for supernatural beliefs come from the different forms of religion – from traditional organized ideologies to various types of New Age mysticism that appeal to gods, angels, demons, ghosts, or spirits. Each of the world's established religions extols beliefs about entities that have supernatural powers. Whether it is priests preaching in pulpits or pagans dancing naked in the woods, all religions include some form of supernatural belief.[9] But you don't have to be religious or spiritual to hold a supersense. For the nonreligious, it can be beliefs about paranormal abilities, psychic powers, telepathy, or any phenomena that defy natural laws. Those who do not pray in temples or churches may prefer to tune in to one of the many cable television channels dedicated to paranormal investigation, or call one of the multitude of psychic telephone networks looking for answers. Even beliefs about plain old luck, fate, and destiny are supported by our supersense. Why else would newspapers print horoscopes if their readers did not pay attention to them? Religion, paranormal activity, and wishful thinking are three points on a continuum of supernatural thinking. You may just entertain one or possibly all three different realms of belief, but they all depend on a supersense that they are real.

The supersense is also behind the strange behaviours or superstitions in which we try to control outcomes through supernatural influence. When a group acts upon these superstitions, we call them ceremonial rituals. When they are personal, we call them individual quirks. Religions are full of rituals to appease the gods, but, outside of the church or temple, there are all sorts of secular rituals that people use to exert control over their lives. These range from the simple superstitions handed down through cultures such as knocking on wood to bizarre idiosyncratic personal rituals we engage in to bring us luck. Even the corridors of power are not free from the supersense. Tony Blair always wore the same pair of shoes in the House of Commons at Prime Minister's Question Time.[10] During his Presidential campaign, President Barack Obama carried a lucky poker chip. He also developed a bizarre superstitious ritual of playing basketball on the morning of every election in his path to the White House. His opponent, John McCain, was open about his catalogue of superstitions, aways carrying a lucky feather and a lucky compass from his Vietnam piloting days. One wonders why, as he was shot down and spent many years as a prisoner of war. During the presidential race, McCain also always carried a lucky penny, a lucky nickel, and a lucky quarter.[11] Apparently, this sum of 31 'super cents' was not enough to secure presidential victory for this luckless senator. When you scratch the surface, you find many of us have a supersense operating beneath the veneer of rationality.

Sometimes our supersense is not even obvious. It can lurk away in the back of our minds whispering doubt and warning us to be careful. It can be that uncomfortable feeling we experience when we enter a room, or the conviction that we are being watched by unseen eyes when no one is there. It can be our unease at touching certain objects or entering certain places that we feel have a connection with somebody bad. It can be the foods and potions we ingest that we think will alter our bodies and minds through magical powers. It can be the simple sentimental value we place on a worthless object that makes it unique and irreplaceable.

SuperSense is about all of the above and more. In this book I expose a wide range of human beliefs and behaviour that go beyond traditional notions of the supernatural. This book is not just about ghosts and ghouls. Rather it is about supernatural thinking and behaviour in everyday human activity. In this way, I hope to show you that we often infer the presence of hidden aspects of reality and base our behaviour on assumptions that would have to be supernatural to be true. Whenever our beliefs appeal to mechanisms and phenomena that go beyond natural understanding, we are entering the territory of supernatural belief. Of course, there are many things we cannot explain, but not understanding them does not make them supernatural. For example, consider a problem we experience every waking moment. How does our mind control our bodies? How can something that has no physical dimensions influence something physical like the body? This is the mind–body problem that we will discuss in chapter 5. Science may not yet understand the mind–body issue and it may never, but that does not make it supernatural because we can investigate the mind with scientific studies to test if the results fit with the predictions.

In contrast, evidence for the supernatural is elusive. When you try to gather evidence for the supernatural, it vanishes into thin air. It is almost always anecdotal, piecemeal, or so weak it barely registers as being really there. Experiments on the supernatural invariably amount to nothing. Otherwise, we would be rewriting the science textbooks with new laws and observations. That's why most conventional scientists do not bother to conduct research on the supernatural. But lack of scientific credibility does little to dent the belief – most of us have a supersense telling us that the evidence is really there and that we should simply ignore the science and keep an open mind. The problem with open minds is that everything falls out – including our reason.

This book is about the science behind our beliefs – not whether these beliefs are true or not. It should change the way you judge other people. When you understand the supersense, you will better understand both your own beliefs and, more importantly, why others hold supernatural beliefs. It should give you insight. It may even make you look at religion and atheism in a new way and realize that everyone is susceptible to supernatural beliefs. I will show that common supernatural beliefs operate in everyday reasoning, no matter how rational and reasoned you think you are. Maybe I should claim that this book will change your life and attitudes towards beliefs but I am not so sure. Because whatever I am about to tell you will go in one ear and out the other. That's the nature of belief. It's really difficult to change with reason. Where does such stubborn thinking come from in the first place?

As part of human culture, we are so immersed in storytelling that it is easy to assume that all beliefs come from other people telling us what to think. This is especially true when it comes to things that we cannot directly see for ourselves. We believe what we are told on the basis of trust. However, this book offers another possible explanation for why we believe in the unbelievable and I think we need to look to children for the answer.

The alternative view for the origin of supernatural beliefs I want to propose is a natural, scientific one based on mind design. By design, I mean a structured organized way of interpreting the world because of the way our brains work. Yes, culture feeds each child with stories but there is more to belief than simply spreading ideas. As the forefather of modern science Francis Bacon said, we prefer to believe what we prefer to be true. I would add that what we believe to be true might come from our way of seeing the world as a child. In other words, the frame of mind within every child leads him or her to believe in the supernatural.

If a supersense is part of our natural way of understanding the world, it will continue to reappear in every child born with this frame of mind. If so, then it seems unlikely that any effort to get rid of supernaturalism will be successful. At the very least, it is going to be a very hard battle to win. It will always be there lingering away in our minds. Even those with a scientific education may still continue to harbour deep-seated childish notions that lie dormant in their adult minds. Should we even try to get rid of them?

SACRED VALUES

The human species may actually need a supersense – not simply because it promises something more than is available in this life, like a security blanket of reassurance for what happens to us when we die, but rather because the supersense enables us to appreciate *sacred values* while we are still alive.[12] We all need sacred values in our lives. Our sacred values can reside in an object, a place, or even a person. We may find the sacred in a word or an act. If you are religious, your world is full of the sacred – places you must go, objects you must revere, individuals you must worship, words you must say, and acts that must follow sacred rituals. But what if you are not religious? Are you immune from sacred values? I am not so sure.

Humans are social animals, and to participate in society we have to share conventions: things that we all agree have some common value. These are the things that can hold a group together. Some conventions are everyday and mundane, such as the money convention of exchanging pieces of paper or lumps of metal for goods. Others are more profound. Certain documents, such as Magna Carta, or the US Declaration of Independence, are more than just pieces of paper. They are sacred objects. They represent important points in civilization, but we revere them as objects in themselves. There's something more to them than simply the words written on them. Or a sacred item could be a book or a painting, a Mozart manuscript or an original Vermeer. Both can be copied and duplicated but it's the originals we value the most. In the same way, a building or location can be sacred. Shrines and churches are obviously holy to the religiously devout, but we can all share in a deeper sense of the value of a place. If you support Manchester United, it's Old Trafford. If you are a Chicago Cubs baseball fan, it's Wrigley Field. These stadiums are more than just sports arenas. To the fan they are hallowed grounds, imbued with as much sacred value as a temple.[13]

Society needs sacred values – anything that we hold to be special and unique beyond any given sum. You can't put a price on a sacred value, or at least you should not willingly do so. Because they cannot be reduced to any scientific or rational analysis, sacred values represent a common set of beliefs that bind together all the members of a group and apply to all of them. Without sacred values, society would deteriorate into a free-for-all in which individuals are only out for themselves. When our societies have sacred values, we are all bound to acknowledge and conform to the group consensus that there are some things that simply should not be bought, owned, or controlled by another group member. Sacred values confirm our willingness to be part of the group and share beliefs even when such beliefs lack good evidence.

Over the coming chapters I hope to show you how our supernatural beliefs can make sense of our sacred values. Don't take my word for it. That would be storytelling. Rather you, the reader, need to come up with your own opinion based on the evidence presented in the following pages. So that you can navigate the path ahead more clearly, let me show you the roadmap.

In the opening chapter, I begin with the notion of 'mind design' – something organized in the way we interpret the world around us – and how it produces some surprising beliefs. Most of us willingly accept that our minds can make mistakes, but we all think we can overcome these errors if given the right information. That's because we all think that we are reasonable. Have you ever heard anyone admit that he or she is unreasonable? Despite our confidence in our own reason, sometimes our capacity to be reasonable is undermined by our gut reactions, which can kick in so fast that it's hard to rein them in with reason. Take the example of evil and our belief that it can be physically real. If you don't believe me, consider how you would feel if you had to shake hands with a mass murderer such as I discuss in chapter 2. Why do we recoil at the thought? Why do we treat their evil as something contagious?

I then want to turn your attention to origins. Tracing the first evidence of supernatural beliefs to the beginnings of culture, I show that, while science has made considerable strides over the last four hundred years, supernaturalism is still very common. Then I want you to consider origins within the individual and the development of belief in the growing child. One of the main points I want to make in the book is that children naturally reason about the unseen aspects of their world, and doing so sometimes leads them to beliefs that form the basis of later adult supernatural notions. In particular, the ways in which young children reason about living things and about what the mind is and can do clearly show the beginnings of ideas that become the basis for adult supernatural beliefs. These are emerging long before children are told what to think, which brings me back to one of the major themes of the book: supernatural beliefs are a product of natural thinking.

Over the next couple of chapters, I examine this natural thinking and how children organize the world into different kinds of categories. In doing so, they must be thinking that the physical world is inhabited by invisible stuff or essences. Science may be able to teach children about real stuff that makes up the world, such as DNA and atoms, but our childish essential reasoning continues to influence the way we reason and behave as adults. This is no more obvious than in the case of our attitudes toward sacred objects. Sacred objects are deemed to be special by virtue of their unique essence, which people believe connects them to significant other people. These can be parents, lovers, pop stars, athletes, kings, or saints – anyone with whom we feel a need to make a connection.

The remaining chapters of the book focus on sentimentality and the irrational fears that we can so easily detect in others but often fail to recognize in our own reasoning. Before concluding the book, I examine the latest thoughts about a brain basis for individual differences in the supersense. Some people are much more willing to entertain supernatural beliefs even when they are highly educated. How can we understand this? Here we consider the brain mechanisms that may be responsible for generating and controlling beliefs and how these can change over the course of a lifetime or during an illness.

By the time you get to the end of this book, I hope you will appreciate that the development of a child's mind into that of an adult is not simply a case of learning more facts about the world. It also involves learning to ignore childish beliefs, which requires mental effort. Education helps, but it's not the whole story. We need to learn to control our childish beliefs. I also briefly consider why there may be a connection between the supersense and creativity. Maybe creativity depends on our capacity to leap over logic and generate new ways of looking at old problems. In which case, creativity and the supersense may be stronger in those of us who are less anchored to reality and more inclined to sense patterns and connections that the rest of us miss or simply dismiss. They are always there in the background of our minds, pushing us toward the supernatural.

In the final pages, I bring these issues together and return to the supersense and the notion of sacred values with an explanation for why human society needs to believe that there are some things in life that must be considered unique and profound. Not only is there room for such beliefs in the modern mind, but they may be unavoidable.

What people choose to do with their beliefs is another matter. Whether religions are good or bad is a heated debate that I will leave to others. I just think that supernatural beliefs are inevitable. At least knowing where they come from and why we have them makes it easier to understand belief in the supernatural as part of being human.

So let's begin that scientific search for the supersense.

CHAPTER ONE

WHAT SECRET DO JOHN MCENROE AND DAVID BECKHAM SHARE?

WEIRD STUFF HAPPENS all the time. Some years ago, before we were married, Kim and I travelled to London. It was our first trip to the capital, and we decided to use the Underground. London's Underground train system transports more than three million passengers every single day, and so we were relieved to find two seats together inside one of the crowded carriages. As we settled down, I looked up to read the various advertisements, as one does to avoid direct eye contact with fellow passengers, but I noted that the young man seated opposite seemed vaguely familiar. I nudged Kim and said that the man looked remarkably like her brother, whom we last heard was travelling in South America. It had been years since we last saw him. Kim stared at the man, and at that instant the man looked up from the paper he was reading and returned the stare. For what seemed a very long time, the two held each other's gaze before the quizzical expression on the man's face turned to a smile and he said, 'Kim?' Brother and sister could not believe their chance encounter.

Most of us have experienced something similar. At dinner parties, guests exchange stories about strange events and coincidences that have happened either to them or, more typically, to someone else they know. They talk about events that are peculiar or seem beyond reasonable explanation. They describe examples of knowing or sensing things either before they happen or over great distances of time and space.

They talk of feeling energies or auras associated with people, places, and things that give them a creepy sensation. They talk about ghosts and sensing the dead. It is precisely because these experiences are so weird that they are brought up in conversation. Pierre Le Loyer captured this notion well four hundred years ago in writing about spirits and the supernatural when he said: 'It is the topic that people most readily discuss and on which they linger the longest because of the abundance of examples, the subject being fine and pleasing and the discussion the least tedious that can be found.'[1]

Most of us have had these bizarre experiences. Have you ever run into a long-lost friend in the most unlikely place? How often have you thought of someone only to receive a phone call from that person out of the blue? Sometimes it seems as if thoughts are physical things that can leap from one mind to another. How often have two people puzzled and said, 'I was just thinking the same thing!' Many of us feel that there is something strange going on. Humans appear synchronized at times, as if they were joined together by invisible bonds. Some of us get a sense that there are mysterious forces operating in the world, acting to connect us together, that cannot be explained away. How do we make sense of all these common experiences?

Many people believe that such occurrences are proof of the supernatural. Beliefs may turn out to be true or false, but supernatural beliefs are special. To be true, they would violate the natural laws that govern our world. Hence, they are *super*natural. For example, I may believe that the British Secret Service murdered Princess Diana in a car crash in Paris. That belief may be true or false. Maybe they did and maybe they did not. It's not impossible. To be true, my belief would have to not violate any natural laws. All that would have been required was a very elaborate plan and cover-up. So it is possible that the British Secret Service murdered Princess Diana – but unlikely. However, if I believe that someone can communicate with the dead princess, then that would be a supernatural belief because it violates our natural understanding of how communication between two people works. They usually both have to be alive. As Michael Shermer says, 'We can all talk to the dead. It's getting them to talk back that's the hard part.'[2]

People can be fully aware that their beliefs are supernatural and yet they continue to believe. Why do people believe in things that go against natural laws? It cannot simply be ignorance.

The answer is evidence. The number-one reason given by people who believe in the supernatural is personal experience.[3] In one survey, half the number of spouses of recently deceased partners reported feeling the presence of the dead.[4] One third reported seeing their ghost. Even my late father-in-law, a brain surgeon of eminent status, saw the ghost of his recently deceased wife, my mother-in-law. Throughout his career he dealt with patients with brain damage and was very familiar with the peculiar experiences the mind can generate. He knew he was hallucinating at the time of his wife's death but that did not stop him seeing her. People have ample opportunity and evidence to draw upon. Of course, other people influence what we think, but firsthand experience gives us a mighty powerful reason to believe. As they say, 'Seeing is believing' and, when it happens to you, it proves what you suspected all along.

For believers, examples of the supernatural are so plentiful and convincing that to simply ignore all the evidence is to bury our heads in the sand. But is there really such an abundance of examples of the supernatural? One major problem is that we are simply not good at estimating the likelihood of how often weird stuff happens. We tend to overestimate the likelihood of events that are very rare, such as being killed in a plane crash. At the same time, we underestimate the likelihood of events that are really quite common. For example, what is the likelihood of two strangers at a party sharing the same birthday? Let's say you're the sociable type and attend a party about once a week. Take a guess at how many people have to be at a party for two of them to share a birthday at half the parties you attend throughout the year. What sort of number do you think you would need? I imagine most of you have come up with quite a big number. But would you believe that statisticians tell us the minimum number is only twenty-three! If you go to a different party each week, with at least twenty-three new people at each, on average two people will have the same birthday half of the time. Or, to put it another way, among the thirty countries taking part in the 2010 World Cup soccer tournament in South Africa, half of the twenty-three-member teams taking part will include two players with the same birthday.[5] What could be more unlikely? Now think of how much more common it is for two people to share the same astrological sign when there are only twelve of those compared to 365 different birthdays in the year. People seem so surprised to meet someone with the same astrological sign and often consider this some sort of fateful coincidence. Our minds are simply not equipped to think about likelihood very accurately, and so we interpret these coincidences as if something supernatural were involved. When we hear of examples that seem bizarre, we treat them as auspicious. The thing about coincidences is that they are not the exception but the rule. As Martin Plimmer and Brian King have observed:

> We frisk each other for links. We're like synchronized swimmers in search of a routine. We relish connections, and we're a highly connected species. If it were possible to map all human activity, drawing lines between friends and relatives, departures and arrivals, messages sent and received, desires and objects, you would soon have a planet-sized tangle of lines, growing ever denser, with trillions of connections.[6]

Uncanny events punctuate our lives, but they seem unusual and beyond explanation. We treat them as significant and profound, leading many of us to believe that there must be supernatural powers at work. Most of us entertain these beliefs even though we may deny them. I am going to show how rational, educated adults as well as the more superstitious among us behave as if there were invisible supernatural forces and energies operating in the world. Over the course of the book, I am going to present a theory that explains why we believe and why some of us are more prone to belief than others. I am going to focus on the individual rather than culture because I think the answer can be found within each one of us.

SOMETHING MORE TO REALITY

The great American philosopher and early psychologist William James wrote more than one hundred years ago that ordinary people tend to believe not only in the reality of existence but in the presence of 'something there' – something intangible that we are bound to infer over and beyond what our normal senses detect:

> But the whole array of our instances leads us to a conclusion something like this: It is as if there were in the human consciousness a sense of reality, a feeling of objective presence, a perception of what we may call 'something there', more deep and more general than any of the special and particular 'senses' by which the current psychology supposes existent realities to be originally revealed.[7]

James is telling us that it is natural to think that there is something more to reality. This something is unknown, unseen, and unmeasurable, and beyond natural explanations. It is supernatural. Moreover, this sense of something more is the basis of all the world's religions, which

> all agree that the 'more' really exists; though some of them hold it to exist in the shape of a personal god or gods, while others are satisfied to conceive it as a stream of ideal tendency embedded in the eternal structures of the world. They all agree, moreover, that it acts as well as exists, and that something is really effected for the better when you throw your life into its hands.[8]

Why do people think like this? Why do we come to believe that there must be something more to nature than can be measured? Where do these ideas come from? From where do we get our supernatural beliefs? There are two schools of thought here: either these are ideas that we hear from other people or they are ideas that partly come from within us. Let's examine both propositions. First, we may be born to believe anything and everything we are told by others. So beliefs are simply the stories we tell each other, and especially our children. Alternatively, we may be born to believe, and what we think might be possible is a reflection of our own way of seeing the world.

Consider the first explanation. Children believe what they are told by adults. We love to tell them about fantasy figures like Santa Claus, the 'Tooth Fairy', and even the 'Bogeyman' if they are misbehaving: 'If you are good, Santa will bring you that PlayStation' or 'If you misbehave, the Bogeyman will take you.' Fairy tales have been around for a long time as a way of teaching our children how to behave. All of the characters in these stories are magical – cats that can talk, witches that can fly, and so on. Characters with supernatural powers are understood to be special and thus are more easily remembered. Because they are so unusual, they work. Isn't it ironic that we immerse our children in make-believe as pre-schoolers, only to tell them to put away such foolish ideas and 'grow up' when they reach school age?

The psychologist Stuart Vyse argues that culture is most important when it comes to the supernatural: 'We are not born knocking on wood; we learn to do so. We are not innate believers in astrology; we become believers.'[9] I agree in part. Many rituals are passed on as customs and traditions. Some of them are so old that we have forgotten why we do them. Every year in the West, children take part in the archaic ceremonies and rituals associated with Halloween and Christmas, mostly unaware of their true origins.[10] On All Hallows' Eve, the practice of dressing up in scary costumes was intended to banish evil village demons. Kissing under the mistletoe and lighting the Yule log were originally pagan fertility rites that became incorporated into Christmas activities. Today we observe these rituals because they have become traditions handed down to us through our culture. But a purely cultural explanation is missing an important point. Why are we so inclined to engage in ceremony and rituals? People may treat these festivals as a bit of fun, but many still believe in real supernatural phenomena. Why would a person accept the supernatural in the first place?

The obvious answer is that there is a real benefit to believing what others tell you. Communicating and sharing ideas with others expands your knowledge so that you don't have to discover everything by yourself. And who best to learn from but older and wiser members of the tribe? If they say that certain plants have healing powers or that some caves are dangerous, it is sensible to believe what they say. In this way, beliefs can easily pass from one generation to the next. If culture and society spread belief, then we should be careful what we tell our children. If this is the root of supernatural thinking, then perhaps we should be held responsible for informing the naive and the young who do not yet know.

This is why the biologist Richard Dawkins thinks that religion is a form of child abuse. He wants a world without God, religion, or any form of supernaturalism. There is only room for science, he asserts, when it comes to understanding nature. Dawkins accuses the churches of indoctrinating our youth with superstitious beliefs. Children are 'information caterpillars' with 'wide open ears and eyes, and gaping, trusting minds for sucking up language and other knowledge'. They gullibly gobble up any facts because of an evolved predisposition to trust whatever their parents and elders tell them.[11]

This brings me to the second explanation for beliefs that I want to draw to your attention. The problem with the gullibility view is that most researchers who study the development of the mind do not regard humans as blank slates for any idea or belief. Rather, the bulk of the work on young children's thinking shows that, before they are capable of instruction, pre-school children are already deeply committed to a number of misconceptions. I think that these misconceptions are the true origin of adults' supernatural beliefs. Yes, culture and church play a role in supernatural belief, but they do not act alone. Rather, they provide the framework to make sense of our own beliefs that we come up with by ourselves.

Even if ideas are transmitted by culture, we still have to answer two fundamental questions: Where did the first supernatural ideas originate? And why do so many isolated cultures share the same basic misconceptions? The common types of belief and reasoning shared by distant cultures, long separated in time and far distant geographically, suggest something intrinsic to the way humans think. For example, almost all cultures have creation myths to explain the origins of the world and the diversity of life that usually involve gods. Gods and spiritual agents are also held responsible for unforeseen events. Whenever we find such universal beliefs and behaviours, we should start looking for reasons why these explanations of origins and events are similar. Like the instinct for language found in every society since the beginnings of civilization, is it possible that a supersense is also part of the human endowment? Do we all start off with a natural inclination to the supernatural that only some of us can overcome? Why is it so damned hard for people to become scientific in their thinking?

I think supernatural beliefs work so well because they seem plausible. And they seem plausible because they fit with what we want to believe and already think is possible. They also make sense of all the weird and uncanny events that pepper our lives. Ideas and beliefs may be transmitted, but only those that resonate with what we think is possible take hold and make sense. This is a really important point that is often overlooked. We either accept ideas or reject them, but seldom do we consider why. Ideas have to fit with what we already know. Otherwise, they do not make sense.

To prove this, let me give you a new idea I want you to believe in. It's not a supernatural one, but it makes the point about how ideas work. If I told you that 'colourless green ideas sleep furiously', would you believe me? Think about it for a moment and try to take the idea on board. At first it sounds okay, but eventually you see that the idea is meaningless. The statement is actually a famous sentence among scientists who study language and thinking. In 1957 the linguist Noam Chomsky constructed this perfectly grammatical but completely meaningless phrase to demonstrate that sentence structure alone is not enough to convey ideas.[12] The content of the sentence follows all the rules of language, but as a sentence it does not compute in our minds. It is meaningless because of what we already know about colour, ideas, sleep, and anger. Something cannot be both green and colourless. Ideas do not sleep. Sleep is not normally furious. These are concepts that already exist in our minds and, because they contradict each other, they dictate that Chomsky's statement makes no sense. So any new idea has to fit into existing frameworks of knowledge. This is why some ideas can be so difficult to grasp. Science, for example, is full of ideas that seem bizarre simply because we are not used to them. It's not that people are being stupid when it comes to science. Rather, many scientific ideas are just too difficult for many of us to get our heads around. On the other hand, folk beliefs about the supernatural seem quite possible. That's why it is easier to imagine a ghost than a light wave made up of photons. We have seen neither, but ghosts seem plausible, whereas the structure of light is not something we can easily consider.[13]

MIND DESIGN

Mind design is the reason why certain ideas are obvious while others are obscure. By mind design I mean the organized way in which our brains are configured to understand and interpret the world. The brain, like every other part of the human body, has evolved over millions of years. Your hands have been designed to manipulate objects. Your legs have been designed for bipedal locomotion. Your liver has been designed to do all sorts of jobs. Likewise, your brain has been designed in certain ways through the process of evolution. Most scientists agree that the brain has many specialized, built-in mechanisms that equip us to process the world of experience. These mechanisms are not learned or taught by others. They form the package of mental tools that each of us is equipped with as part of our mind design. But this design does not need a designer. You don't need a god to explain where the design came from. It's simply the way gradual adaptation of biological systems through the process of evolution has produced a complex problem-solver. Natural selection is our designer.

The brain did not fall out of the sky, ready packaged to deal with the world.[14] Rather, our brains gradually evolved to solve the problems that faced our ancestors. Our complex modern brain has emerged by accumulating small, subtle changes in its structure passed on from one generation to the next. This is the field of evolutionary psychology, and, as the computer scientist Marvin Minsky succinctly puts it, the mind is what the brain does. Our minds are constantly active, trying to make sense of the world by figuring out how it works. This is because the world is complex, confusing, and filled with missing information. Each of us is a sleuth trying to complete the puzzle, find the culprit, and solve the crime when it comes to understanding.

What we do naturally and spontaneously at the most basic level is look constantly for patterns, imagining hidden forces and causes. Even the way we see the world is organized by brain mechanisms looking for patterns. At the turn of the twentieth century, the German Gestalt psychologists demonstrated that humans naturally see patterns by organizing input with certain unlearned rules. What these early psychologists realized was that the world is full of input that is often cluttered, ambiguous, or simply missing. The only way the mind can sort out this mess is by making guesses about what is really out there.

For example, a pattern made up of four pies with a slice taken out of each one is usually seen as a white square sitting in front of four dark circles. Our mind even fills in the missing edges of the square in between the pies. But the square does not really exist. Our brains have created something out of nothing. More spookily, we can measure activity in those areas of the brain that would be active if the square really existed! This area, known as the visual cortex, is a three-millimetre layer about the size of your credit card that sits directly at the back of your head. Contrary to popular misconception, it's not your eyes but your brain that does the seeing. The brain cells in this region are all related to vision in some form or another. So, in this region, the brain registers what is *really* out there in the world, makes a decision about what *should* be out there, and then generates its own brain activity *as if* what it has decided should be out there really is.[15] Even when a perception is a trick of the mind, it still shows up as real brain activity. This filling-in process reveals how our brains are wired to make sense of missing information. Four-month-old babies also see this ghostly square.[16] We know this from a simple behaviour: babies get bored when shown the same pattern over and over again. Wouldn't you? So if you present babies with the ghostly square, they eventually stop looking at it. If you then show them a real square, they remain bored, whereas they perk up and get excited if you show them something else, like a circle. In other words, they must have seen the illusory square, eventually got tired of looking at it, and found the real square just the same as the imaginary one their mind had created out of nothing. Such studies tell us that baby brains are designed for filling in missing information and making sense of the world.

As my colleague Richard Gregory has argued, illusions like the missing-square pattern reveal that the mind is not lazy. Our minds are actively trying to make sense of the world by thinking of the best explanation. For example, if someone took a handful of coffee beans and scattered them across a table in front of you, you would immediately see patterns. Some beans would instantly cluster together into groups as you simply looked at the array. Have you ever watched the clouds on a summer's day turn into faces and animals? You can't stop yourself because your mind has evolved to organize and see structure. The ease with which we see faces in particular has led to the idea that we are inclined to see supernatural characters at the drop of a hat. Each year some bagel, muffin, burnt toast, potato chip, or even ultrasound of a

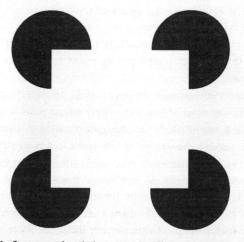

FIG. 2: Both infants and adults see an illusory white square in the typical Kaniza figure. AUTHOR'S COLLECTION.

fetus showing the face of some deity is paraded as evidence for divine miracles.

We also seek out patterns in events. Our mind design forces us to see organization where there may be none. When something unusual or unexpected happens, we immediately look for order and causes. We cannot handle the possibility that things happen randomly by chance. It may even be impossible for the mind to think in terms of random patterns or events. If I asked you to generate a random pattern, you would find this incredibly hard to do. Try it out for yourself at a keyboard. Empty your mind and simply press either the '1' or '0' key whenever you feel like it. Be as random as you can. For example, here's my attempt with forty-eight key presses:

1 0 0 1 1 0 0 1 0 1 0 0 0 1 1 1 0 0 1 0 0 1 0 1 1 0 1 1 0 0 1 0
1 0 1 1 0 1 0 0 1 0 1 1 0 0 1 1

I felt I was being random, and at first glance the pattern looks pretty disorganized. If you count the number of times I typed '1', then I have done pretty well with exactly half (twenty-four). Now consider the same sequential key presses in groups of two.

10 01 10 01 01 00 01 11 00 10 01 01 10 11 00 10 10 11 01 00
10 11 00 11

There are five 00 pairs, seven 01 pairs, seven 10 pairs, and five 11 pairs. If the sequence was truly random, then these pairs should be equal, but I was much more likely to alternate (fourteen times with either 10 or 01) key presses than to press the same key twice (ten times with either 00 or 11). The difference may seem small, but it becomes highly significant over more trials. If you break the sequence into the eight possible triplets, then the patterns become even more obvious.

Our brain has its natural rhythms that it likes to settle into. This is how the best rock–paper–scissors players succeed. To remind you, it's a game between two players in which, after the count of three, each player has to produce a rock (fist), paper (open hand), or scissors (first two fingers open). Scissors beats paper, which beats rock, which in turn beats scissors. The object of the game is to guess what your opponent will produce. To succeed you have to be as random in the three options as possible. World champion players (yes, they do exist) are not psychic.[17] They are expert at detecting patterns and generating their own random sequences, but this skill requires a lot of mental energy, especially from the frontal parts of the brain that control planning.[18]

It is just as difficult to think and act randomly by effort of will as it is to perceive a random world. Because our minds are designed to see the world as organized, we often detect patterns that are not really present. This is particularly true if we believe that patterns should be there in the first place. So someone who believes that supernatural forces operate in the world is on the lookout for examples of strange, inexplicable phenomena and conveniently ignores the multitude of mundane events that do not fit this interpretation. We forget every typical phone call but remember the unexpected one because it draws our attention. The flip side of mind design is that we also fail to realize that events that we think are highly unlikely are in fact not so unlikely. Meeting people at a party who share the same birthday seems unlikely. With this bias toward detecting patterns, someone who is inclined to supernatural belief has ample opportunity to see evidence for significant chains of events where there is none. This is the product of our mind design, and there is good evidence that we all differ in the extent to which we see order or chaos in the world. Later, I examine the idea that the difference between believers and nonbelievers may be due more to how they interpret the world than to what they have been told to believe.

In addition to organizing the world into patterns, mind design leads us to seek deeper, hidden causes operating in the world. Much of what controls the world is hidden from direct view, and so our minds have evolved to infer the existence of things we cannot see. We try hard to understand outcomes of events that have already happened and to which we were not privy. For example, imagine you arrive home to find a plate broken on the kitchen floor. *How did this happen?* you ask yourself. You start to reconstruct the order of events. The plate was on the table when you left that morning. Has someone else been in the house? Has there been an earthquake? Like a detective, you work backwards in time trying to reconstruct why something happened. This is how we interpret and understand a chain of events. However, such reasoning can also lead to mistakes. A human mind that links events in this way is always in danger of committing the mistake of *post hoc, ergo propter hoc*: 'after this, therefore because of this', which means that we tend to group events together in a causal way. We see the first event as having caused the second. There are two problems with this. First, we infer the actions of forces where there may be none, and second, we tend to link events that are not actually even related.

By linking events together, we see sequences in terms of cause and effect. For example, consider a very simple event involving objects colliding with each other. Imagine watching a game of billiards or pool. If we see a white ball strike a red ball, we see one event causing another. It's the same for babies. If you show seven-month-old babies similar collision events, they interpret the first ball as causing the second to move, because if you reverse the sequence, they treat the reverse event as something different.[19] Like adults, they see the red ball launching the white one. Nothing odd here you might think. In fact, you might say this is a very sensible way to interpret the world. However, the seventeenth-century Scottish philosopher David Hume tells us that such intuitions are an illusion because you cannot directly see cause. You cannot actually see the forces at work. You only see one event and then another event. This may seem far-fetched until you consider cartoon animations. When we observe a cartoon ball striking another, we infer the same causal force, but of course there is none. A cartoon is simply a set of drawings. Our mind interprets the sequence as if one ball were colliding with another. It is an illusion that helps us understand the world in terms of real forces because we often do not or cannot observe them at work.

So your mind design forces you to see patterns and to think something caused the patterns to form. You infer that what may be completely unrelated events are connected in some way. Things that happen after each other appear to be caused by forces that may not exist. This is all the more true when the outcome is not predictable, as in a game of chance. When something unexpected happens, you instinctively look for whatever caused it to happen. This type of thinking explains superstitious behaviour: repeating actions or engaging in certain behaviours in an effort to control outcomes. For example, if you have a particularly successful day on the tennis court or at the poker table, you may feel a strong compulsion to duplicate whatever actions you took that day in an effort to repeat the success. It may be wearing a particular piece of clothing or sitting in a favorite seat. Soon these behaviours may become essential routines and obsessions.

Athletes are notorious for their superstitious rituals.[20] Rituals usually start off as innocent habits – something we all have – but because they become linked to important outcomes (like winning a game), they can take over an individual's life. The tennis ace Jelena Dokic was probably the most complicated in her rituals, or at least the most honest and open about them. First, she avoided standing on the white lines on court. (John McEnroe did the same.) She preferred to sit to the left of the umpire. Before her first serve she bounced the ball five times, and before her second serve she bounced it twice. While waiting for serves, she would blow on her right hand. The ball boys and girls always had to pass the ball to her with an underarm throw. Dokic made sure she never read the drawsheet more than one round at a time. Finally – and bear this in mind sports memorabilia collectors – she always wore the same clothes throughout a tournament. Pheweee!

Jelena is not alone. Every year when I monitor exams I see a number of intelligent young adults engaging in routines (one had to walk around her table three times) or producing a multitude of lucky charms and 'gonks' (troll-shaped soft toys) that they believe will improve their performance. Even if you don't believe in these rituals and charms, what's the harm in trying? Well, none, unless they take over your life and prevent you from achieving your goals, as illustrated by Neil the Hippie from the 1980s' UK comedy about student life, *The Young Ones*:

> I sat in the big hall and put my packet of Polos on the desk. And my spare pencil and my support gonk. And my chewing gum

and my extra pen. And my extra Polos and my lucky gonk. And my pencil sharpener shaped like a cream cracker. And three more gonks with a packet of Polos each. And lead for my retractable pencil. And my retractable pencil. And spare lead for my retractable pencil. And chewing gum and pencils and pens and more gonks, and the guy said, 'Stop writing, please.'[21]

Superstitions are common in situations where the factors that control outcomes are unpredictable or the consequences of something going wrong could be fatal. However, rituals are also common among many high-achieving individuals in situations where attention to detail can lead to success. Harrison Ford, Woody Allen, Michelle Pfeiffer, and Winona Ryder are just a few celebrities who allegedly engage in ritualistic behaviour. In a recent TV interview, the soccer star David Beckham described some of his unusual rituals:

I have got this disorder where I have to have everything in a straight line or everything has to be in pairs. I'll put my Pepsi cans in the fridge and if there's one too many then I'll put it in another cupboard somewhere. I'll go into a hotel room and before I can relax I have to move all the leaflets and all the books and put them in a drawer.[22]

Such behaviours reflect an obsessive attention to detail. It may be the case that those with a personality characterized by a need for discipline and control are more likely to achieve professional success in their striving for perfection. Such individuals can be found in all walks of life. We all know people who seem to pay excessive attention to detail and order. In about two out of every one hundred members of the general public, ritualistic behaviour that controls the individual's life becomes the medical problem of obsessive–compulsive disorder. These sufferers have to engage in ritualistic behaviour and are incapable of breaking out of their routines. They are aware that their behaviours are odd, but that knowledge does not help. The irony is that, if prevented somehow from performing their rituals, they might not perform as well because of their increased anxiety that they are now luckless. These rituals give a sense of control in situations where control is important. So those with obsessive–compulsive disorder are not necessarily irrational, since this 'illusion of control' is psychologically comforting in comparison to no control at all.[23]

However, the belief that rituals work is supernatural. We may deny that rituals are based on supernatural beliefs and claim that many of them, such as throwing salt over one's shoulder when it is spilled on the table, are no more than harmless traditional customs of long-forgotten origin, much like the Christmas rituals discussed earlier. But if we think there is nothing to them, why do we see an increase in such behaviour at times of crisis? During the first Iraq war in 1991, Saddam Hussein fired SCUD missiles indiscriminately into Tel Aviv. What could be more stressful than sheltering during an air raid, not knowing if your family is about to be killed? In subsequent interviews, those living in the highest-risk areas were asked about their experiences, and it was observed that during the conversation they 'knocked on wood' significantly more than those from low-risk areas. It's not clear where the practice of rapping on wood to ward off bad luck first came from. It may be linked to the pagan practice of tapping on trees to signal one's presence to the wood spirits, or maybe it's a reference to the Christian cross. Who knows? Whatever its origin, the threat of danger triggered a superstitious behaviour.[24] We may deny the supersense, but it nevertheless lingers in the background of our minds, waiting for an opportunity to make a guest appearance at times of stress, when rationality can so easily abandon us.

The beliefs behind superstitious practices may be supernatural, but here's the interesting point: they do work to reduce the stress caused by uncertainty. Rituals produce a sense of control, or at least the belief that we have control even when we don't. The illusion of control is an immensely powerful mechanism to immunize against harm, especially if it is unpredictable. Not only do we find it hard to think randomly, but we don't like unpredictable punishment. We all know what it's like waiting for something bad to happen. We just want to get it over and done with as soon as possible. As a child growing up in Scotland, I remember sitting outside the headmaster's office waiting to be 'strapped' for fighting in the playground. I think it was my foreign accent that made me the focus of attention. By that age, stories about the Bogeyman were no longer effective, and corporal punishment was deemed the best deterrent. The strap was a barbaric leather belt specifically designed for whipping the hands – a practice that has now been outlawed. It wasn't the pain of being strapped that was unbearable, however, so much as the wait and the sense of helplessness. I had no control over the situation. Studies of pain thresholds reveal that humans can tolerate much higher electric shocks if they think that they can stop the punishment at any point in comparison to those who do not think they have this option.[25] Doing something, or believing that you can do something, makes the unpleasant more bearable. Without the perception of control, we are vulnerable to our supersense. When adults were asked to think back to a situation when they were without control, researchers discovered that participants were much more likely to see patterns in random pictures, to infer connections between events that happened by chance and even to believe that superstitious rituals were effective.[26] In the absence of perceived control, people become susceptible to detecting patterns in an effort to regain some sense of organization. No wonder those stock market traders are clutching their 'lucky' rabbit's feet as we feel the full brunt of the current economic world recession. 'Doing nothing' is not an option. Anything that we feel can affect outcome is better than nothing, because an inability to act is so psychologically distressing.

It is not just superstitious routines that reinforce the illusion of control. For many, this illusion explains the power of the mind and wishful thinking. The Harvard psychologist Dan Wegner has shown that the same causal mechanism can lead to 'apparent mental causation': an individual's belief that his or her thoughts have caused things to happen when they are closely connected in time. Imagine that you wish someone harm and something bad actually happens to that person shortly afterward. Such a coincidence must occur regularly, but it is very hard not to think that you are responsible in some way. Wegner and his colleagues found that subjects who thought ill of someone behaving like a jerk believed that they had caused his subsequent headache. In fact, the 'jerk' was the experimenters' confederate, and the setup was a scam. Nevertheless, adults readily linked these two events together as if they had cursed the 'victim.'[27] This is all the more apparent in young children, who still are not sure about the difference between mental thoughts and actions. They think that wishing can cause things to actually happen. However, Wegner's research indicates that many adults continue to harbour such misconceptions even though they know that they should not think like this. For example, in games of chance such as gambling, people behave as if they have control when they don't. They feel more confident about winning if they get to throw the dice. They prefer to bet before the dice are thrown rather than after. They think they are more likely to win the lottery if they choose the numbers, and so on. Such behaviour would be utterly absurd if deep down we did not think that we have some influence over events. This is because of our mind design.

Later on, I examine how mind design emerges early in development as children come to understand and predict the physical world, the living world, and the mental world. We will look at studies that prove they must be reasoning about the hidden properties of objects, living things, and their own minds as well as those of other people. I show that young children are thinking about gravity, DNA, and consciousness – all invisible to the naked eye – and that they do this long before teachers have had a chance to fill their heads with ideas. I show that this way of reasoning is very powerful for children's understanding, but that it can also let them down, because reasoning this way about the unseen properties of the natural world sometimes leads to supernatural explanations. Children may learn when they grow up that such supernatural notions are wrong, but what if such childish ideas never really go away?

Most adults believe that when they learn something new that contradicts what they previously thought, they abandon their earlier misconceptions and mistaken ideas. However, it is not clear that this happens entirely: childish notions can linger on in the mature mind. Consider an example from the world of objects. Imagine two cannonballs of exactly the same size. One is made of light wood and the other one is solid iron that is one hundred times heavier. If you were to drop them both at the same time from the leaning Tower of Pisa, what would happen?[28] Children think that heavier objects fall much faster than lighter ones. Heavier objects do land before lighter ones, but only just, and that's because of air resistance. If you dropped the cannonballs in a vacuum where there was no air resistance, they would land exactly at the same time. As a child, I did not believe this until a physics teacher demonstrated that a feather and a coin fall at exactly the same speed in a vacuum. Most college students make the same mistake.[29] The amazing thing is not that adult students get it wrong, but rather that these are students who have been taught Newton's Laws of Object Motion and should know better. They should know the correct answer. Somehow the scientific knowledge they have so painstakingly learned loses out to their natural intuition about weight and falling objects.

The example of the falling cannonballs is important because it reveals that we may never truly abandon our childhood misconceptions when we become adults and learn new facts about the world. Some of us are more vulnerable to these misconceptions than others. Now imagine how difficult it is for us to abandon beliefs that include the supernatural. Here there is precious little evidence to dissuade us of our beliefs. If we hold childish notions about the unseen mechanisms of reality, then the difference between believers and nonbelievers may have less to do with what we have been told and more to do with our susceptibility to our own childish misconceptions. If you are someone who is inclined to believe that there are supernatural forces operating in the world, then you will interpret all manner of events in light of this way of thinking. There will be no chance occurrences. Fate and luck will explain why things happen. You will infer the presence of supernatural agents, and evil and good will become tangible forces.

WHAT NEXT?

Our lives are punctuated by bizarre occurrences. How do we make sense of them? All too often we appeal to explanations that evoke some supernatural activity even though the evidence for such activity cannot be directly observed or studied. So we are left with belief. Where do these beliefs come from? One account is based on the idea that supernatural beliefs are spread by what other people tell us. Certainly this may be true for the content of a belief – the name of a spirit or the nature of the rituals that need to be performed – but what about the basis of the belief? And why are so many of us so willingly gullible? One reason may be that it is our natural way of thinking to assume that there is a supernatural dimension to reality – the 'something there' that William James talked about.

Religion is the most familiar face of such supernatural belief: most religions have deities and other supernatural beings that are not restricted to natural laws. Even many people who do not believe in God are nevertheless willing to entertain the notion that there are phenomena, patterns, energies, and forces operating in the world that cannot be explained by natural laws. God may require supernatural belief, but supernatural beliefs do not require God.

In the next chapter, I want to develop this idea further by demonstrating that most of us can hold supernatural beliefs even when we are not fully aware that we do.

And, for that, I need an old cardigan.

CHAPTER TWO

COULD YOU WEAR A KILLER'S CARDIGAN?

WHEN IT COMES to making choices, most of us feel confident that we evaluate the evidence objectively, weigh the pros and cons, and act according to reason. Otherwise, we would have to concede that our decisions are unreasonable, and few individuals are willing to acknowledge this. But the truth is that human psychology is littered with many examples of faulty reasoning. This is why scientists are so interested in studying the mistakes we make and our biases and logical errors. They seem to fly in the face of reason and suggest that there must be underlying mechanisms responsible for controlling our thought processes. This is the mind design that I talked about in the last chapter. The aspect of mind design that interests me is the one that leads us to infer the presence of patterns, forces, and energies operating in the world where there may be none. This is what I mean by a supersense. Even if you deny having a supersense, you may still be susceptible to its influence, because the processes that lead to supernatural thinking are not necessarily under conscious or willful control. And, as you will see later in the book, some researchers even question whether there is such a thing as conscious willful control.

I like to illustrate this point in the public lectures I give on the origins of supernatural thinking by talking about our reactions to memorabilia. These objects are the best examples because most audiences immediately recognize what I am talking about when it comes to considering the hidden power of simple inanimate objects. To demonstrate the psychological impression created by objects I hand out a black fountain pen dating from the 1930s that once belonged to Albert Einstein. Okay, I lie to the audience about the provenance of the pen, but the belief is sufficient. The reverence and awe towards this object is palpable. Everyone wants to hold it. Touching the pen makes them feel good. Then I ask the audience if they would be willing to wear the cardigan I brought along. Given the oddity of the question and the tattered state of the woollen garment, the audience is understandably suspicious. After a moment's consideration, usually around one-third of the audience raise their hand. So I offer a prize. More hands are raised. I then tell them about Cromwell Street as an image of Fred West rises menacingly from the bottom of the PowerPoint display. Once they are told that the cardigan belonged to Fred West, most hands usually shoot down, followed by a ripple of nervous laughter. People recognize that their change of heart reflects something odd.

There are always the exceptions, of course. Some people resolutely keep their hand raised. Typically, they are male and determined to demonstrate their rational control. Or they suspect, rightly, that I was lying about the owner of the cardigan. What is remarkable is that other audience members sitting next to one of these individuals visibly recoil from their neighbour who is willing to wear a killer's cardigan. How could someone even consider touching such an appalling garment? It's a stunt, of course – a deliberate ploy set up to create a sense of revulsion in an unsuspecting audience.

Last year, this stunt earned me some notoriety in Norwich, England.[1] I was presenting my theory on the origin of the supersense and why science and rationality will not get people to abandon such beliefs easily. The presentation took place at a major British science festival, and the world's science press was there. Since every quality paper had a science correspondent present, I circulated an article outlining my ideas so that there would be a good turnout at the press conference. I argued that humans are born with brains that infer hidden forces and structures in the real world, and that some of these inferences naturally lead us to believe in the supernatural. Therefore, we cannot put sole responsibility for spreading supernatural belief on religions and cultures, which simply capitalize on our supersense.

The cardigan demonstration was meant to illustrate to an educated, intelligent, rational audience (albeit one that included journalists, who are always looking for a 'hook') that sometimes our beliefs can be truly supernatural but have nothing to do with religious indoctrination. Even atheists tend to show revulsion at the idea of touching Fred West's cardigan. If it's true that our beliefs can be supernatural but unconnected to religion, then it must also be true that humans will not necessarily evolve into a rational species, because a mind designed for generating natural explanations also generates supernatural ones.

News of the cardigan stunt and my comments spread like a virus across the world's digital networks. I gave interview after interview, and the event generated web postings on both religious and secular sites with a mixture of ridicule and praise. Some colleagues didn't like the showmanship, but I had made a point that got people talking. People were infuriated. I had touched a raw nerve. It was a sacrilegious stunt, even though no particular religion had been offended. But what had I demonstrated that upset the public so much? What did wearing a killer's cardigan really show? Was it a demonstration of irrationality? How did this prove that humans will not evolve a rational mind?

I think the killer's cardigan illustrates our common supersense. It says something about the sacred values of the group. It also says something about us as both individuals and group members. Their revulsion to the cardigan could reflect a common supernatural belief that invisible essences can contaminate the world and connect us together, almost like some form of human glue. Or at least it feels as if there is something tangible that joins us together. In academic social psychology, 'social glue' is the term to describe the mechanisms for the social connectedness of a group.[2] Any behaviour that causes members of a group to feel more connected can operate as social glue. This is conspicuous at sporting events where many different fans from all walks of life come together as one. Hundreds of complete strangers who would normally not interact with each other suddenly become a highly organized and unified collective. In 1896 the French sociologist Gustave Le Bon described this phenomenon of crowds: 'Sentiments, emotions, and ideas possess in crowds a contagious power as intense as that of microbes.' It is indeed as though something physical infects such groups. Unfortunately for English soccer, very often the power of this mass mentality can overwhelm normally law-abiding individuals who find themselves caught up in hooliganism and brawling with rival teams. Le Bon argued over one hundred years ago that social glue explains why supporters do not feel individually responsible for their actions and claim that they simply went with the crowd.[3]

I see this glue also operating at the individual level. Each of us can feel a special, intimate connectedness to another individual. I believe this mechanism can work at the level of a perceived inner essence. An essence is an underlying, invisible property that defines the true nature of something. It doesn't really exist, but we think and behave as if there were some inner stuff inside people that makes them who they are. I examine this notion more thoroughly throughout the book because it explains a lot of our peculiar behaviour toward others and their possessions. I examine the recent research on essentialist thinking in children and show that this type of thinking can lead us not only to perceive an invisible property that inhabits individuals but also to transfer that property to their objects. It might be natural to believe there is an essential connection or glue that can bind us to others or repel us from them, even though such a connection would be supernatural. That's why I think the cardigan stunt revealed that some people believed that the essence of Fred West had contaminated his clothing.

This essential glue could provide a useful heuristic for interacting with others. Heuristics are simple shortcuts in reasoning that lend support for more complex decision-making processes. We use them all the time when judging other people. Have you ever taken an instant dislike to someone? What was the reason? Often you couldn't say – it was just a feeling you got. When we meet someone for the first time, there's a great deal of unconscious decision-making going on. *Who is this person? What do I know about him? What do I feel about him?* We may be able to reflect on some or all of these questions, but often we answer without being aware of doing so. We are relying on unconscious inferences and heuristics. Social psychologists have shown that, with the barest information, people can make judgements about others rapidly and effortlessly. And yet such fleeting impressions, or thin slicing, as it is known, can have a profound effect on our decisions. For example, students can accurately predict teaching evaluation scores for a lecturer based on as little as two seconds of silent video taken from one of the lecturer's classes. They can even predict which surgeons will be sued for malpractice based on a couple of seconds of muffled speech. Something in the quality of the movements and sounds reveals surprisingly rich information about their social skills.[4] Humans are exquisitely sensitive to judging others, even though we are often unable to say exactly what it is about them we are noticing.

INTUITIVE REASONING

Such unconscious thinking forms part of what I call intuitive reasoning, which to most educated ears sounds like an oxymoron. How can reasoning be intuitive? By intuitive, I mean unlearned. As we shall see later in the book, there is good evidence that children naturally and spontaneously think about the unseen properties that govern the world. They infer forces to explain events they cannot directly see, understand that living things have a life force, and reason in terms of essence when thinking about the true nature of animals. And, of course, they begin to understand that other people have minds. These processes are not taught to children. They are reasoning, though it is not clear that they can necessarily reflect on why or how they are coming up with their decisions. That's why their reasoning is intuitive.

Intuition is often called a 'gut feeling'. Sometimes we get a 'vibe' when we sense a physical feeling of knowing – like the 1960s' hippies, whose talk of getting good or bad vibes was a shorthand for gut feelings. The neuroscientist Antonio Damasio calls this the somatic marker: it indicates the way emotions affect reasoning in a rapid and often unconscious way. 'Somatic' is derived from the Greek word for 'of the body.'. In his remarkable research, Damasio and his wife Hanna have shown that reasoning works by combining information from past experience and encounters and feeding that into decision-making related to the current situation. Past learning is stored as a response deep in the emotional centres of the brain known as the limbic system. Sometimes referred to as the 'reptilian' part of brain because of our shared evolutionary history with reptiles, these centres relay signals into the frontal lobe areas that are concerned with decision-making. If part of this circuit is disrupted through injury, reasoning can be impaired. In one study, patients with damage to their frontal lobes took part in a gambling experiment in which they had to select cards from one of four different decks. Two of the decks paid out low amounts, and the other two paid out greater sums. However, unbeknownst to the players, there were more penalty cards in the high-reward decks compared to the low-reward decks. The frontal lobe-damaged patients were much poorer at learning to avoid the risky decks compared to normal players.

Normally when faced with risk, we sweat. It's a telltale sign of emotion. To understand the role of the emotions in the learning involved in the gambling experiment, the Damasios measured how much sweat each player produced by placing electrodes on their skin. This measure, known as the galvanic skin response, detects changes in skin conductance as a measure of underlying arousal. It's the same principle used in lie detectors. What they found was astonishing. Both normal and frontal lobe-damaged patients showed the same skin conductance before each card was turned over at the beginning of the game. However, as the game progressed and normal players began to learn that some decks were more risky, they became more aroused just before they chose a card from these piles. They were starting to sense the patterns. Bells and lights were going off in their emotional systems to warn them that their decision was wrong. This happened before they were even consciously aware that the odds were stacked against them. Intuition was telling them to be careful. More remarkably, frontal lobe-damaged patients showed no anticipatory arousal whatsoever! Past experience and learning may be vague and unconscious, but they provide a 'feels right' marker that enables individuals to be sure about their decisions. In the Damasios' study, the frontal lobe-damaged patients, who didn't have these markers, were either paralyzed with indecision when having to make a choice or completely careless and unconcerned about the consequences of their actions. This was because they had no somatic marker to help them decide or to warn them to be more careful. They could not feel the answer.[5]

The Fred West cardigan stunt dramatically revealed that my listeners' rapid and automatic intuition kicked in before they had time to consider why they would not wear it. Sadistic killers disgust most of us and, without even thinking about it, we would not want to come into physical contact with them or their possessions. Not all of us, however, feel this way. Psychopaths and sociopaths do not feel any connection with their fellow humans, and that's what enables them to do the inhuman things they do. They don't show the same emotional arousal the rest of us have.[6] However, not everyone who could wear the cardigan is psychotic. Some are simply not sentimental about objects. They may decline the invitation to wear the cardigan, but only because they do not want to stand out from the crowd. Whether we feel the presence of Fred West or simply do not want to be seen to be different, most of us refuse the invitation. Anyone who boldly insists on wearing the cardigan can argue the illogical nature of the association, but that person is still going to lose friends. Would you associate with someone who was not bothered about doing something that most others find repugnant?

I think the main reason the stunt annoyed critics who read about the event was that they probably experienced the same clash between intuition and logic that my audience felt. They initially considered how they would have responded using their inituitive processes, and then, with their rational mind, they realized the logical inconsistency of either a yes or no answer. Also, there is simply no correct answer to the question, making it all the more vexing. Would you wear a killer's cardigan for £1? What about £10,000? There is a point at which most people would change their mind, but what is so undesirable in the first place about touching items owned by evil people or living in houses where murders were committed? Why do the majority of us have these reservations?

The idea for the Fred West cardigan stunt came from the work of Paul Rozin at the University of Pennsylvania.[7] Rozin's experiments are some of the most interesting and provocative examples of the peculiar nature of human reasoning. Much of his work concerns the complex human behaviour of disgust. Disgust is a universal human reaction triggered by certain experiences that elicit a strong bodily response. Anyone can recognize that nose-wrinkling, revolted, nauseous, retching, stomach-churning sensation we get when we are disgusted. It's a powerful and involuntary response that can be difficult to control.

Disgust is interesting because we all develop nausea reactions to specific things such as human excrement and putrid corpses. However, there is also room for learning: certain substances and behaviours can be deemed disgusting if others say so. The diversity of food preferences, personal hygiene, and sexual practices across different cultures proves this. It is well known that Asian cuisine includes insects and reptiles that are considered unpalatable by Western standards. Less well known is the beverage Kopi Luwak, a rare gourmet coffee from Indonesia that is made from beans passed through the digestive system of a palm civet, a dark brown, tree-dwelling, catlike creature found throughout Southeast Asia. Kopi Luwak is sold mainly to the Japanese at up to £300 a pound, making it the world's most expensive 'crappacino'. Or take phlegm. There are few things more revolting than someone else's creamy mucous. In the run-up to the 2008 Beijing Olympics, the city officials tried to outlaw the commonly accepted Chinese practice of clearing phlegm in public by spitting and sinus-clearing, which are nauseating practices to most Westerners. Ironically, the Western practice of blowing the contents of one's nose into a handkerchief and putting the handkerchief, with its contents, in one's pocket could bring on dry-heaves in many Japanese, who consider it disgusting to carry such fluids around on one's person. I guess we in the West would think the same of other bodily excrements kept in our pockets. Or consider sex with animals. Like many others, I had thought bestiality to be universally taboo until I discovered that intercourse with donkeys is acceptable in the northern Colombian town of San Antero, where adolescent boys are actively encouraged in this practice. They even have a festival to celebrate this bestiality; particularly attractive donkeys are paraded in wigs and makeup.[8] (I am still hoping that this last example is an elaborate hoax.)

There's a saying in the north of England, 'There's nowt so queer as folk', and these few examples demonstrate how society and culture can shape what we find disgusting and what we find acceptable. In later chapters, we will see that all of us experience feelings of disgust. Our responses to some disgusting things are automatic and largely unlearned, but the people around us shape others, such as the violation of taboos. In this way, gut-wrenching disgust can be triggered to prevent behaviours that threaten the sacred values of our society.

WHY DON'T WE WANT TO WEAR THE CARDIGAN?

Rozin's work on contamination shows that adults do not want to come into physical contact with disgusting items even after they have been washed. (One of the items he used was Hitler's sweater. It didn't take much ingenuity to adapt this to Fred West's cardigan for a modern audience, as the principles are the same.) Rozin identified at least four reasons why we refuse to touch evil items, and he found that adults endorse each of these reasons to varying degrees.

1. We do not want to be seen to undertake an action that the majority would avoid.
2. Any item associated with a killer is negative, and thus wearing it produces associations with the act of killing.
3. We believe that there is a physical contamination of the clothing.
4. We believe that there is a spiritual contamination of the clothing.

Social conformity, the first explanation, is sensible, but only when we imagine what others would think of us. In other words, why does society regard touching certain items of clothing as so unacceptable?

Why is the physical contact more wrong than simply saying the name or painting a picture of the culprit? The answer lies somewhere in the remaining three reasons.

Many Web critics argued that my cardigan stunt demonstrated simple association only and that there was no need to talk about contamination. However, an explanation based on association rings hollow to my ears. How and why should a cardigan come to represent the negative association with a killer? If I had chosen a knife or noose, the association account would have been adequate. A cardigan is not an item usually linked to murderers. It is something that offers warmth and comfort and, most importantly for my demonstration, intimacy. This combination was meant to jar and shock. The infamous photo of a snarling Fred West taken at his arrest produces a strong association, but personal items such as clothing trigger stronger negative responses. Images are powerful, but objects are even more so. Intimate clothing is more powerful still. That's why you never see someone else's underwear for sale in second-hand clothing stores no matter how well they might have been cleaned and sterilized. This is what Rozin has shown in many similar experiments in which he presents adult subjects with items that have been contaminated. In spite of efforts to sterilize the items, adults still feel disgusted. Something persists in the clothing. More people would rather wear a cardigan that has been dropped in dog faeces and then washed than one that has also been cleaned but worn by a murderer.

What about the explanation based on physical contact? It goes without saying that no one wants to get really close to a serial killer. Maybe you fear for your life, but it could also be that we treat evil as a physical contaminant that could be transmitted by touch. Not touching something contaminated by evil could be another heuristic to avoid having bad things happen to us. Perhaps Mother Nature provided us with a quick and easy rule of thumb: 'If something is bad, do not touch. You might catch it too.' After all, we don't know why someone becomes a psychotic killer. It could be that something they touched or ate drove them mad. In September 2000, twenty-three-year-old Jacob Sexton murdered a Japanese female exchange student in Vermont following a two-month binge on LSD.[9] After beating the girl to death with his bare hands, he lay down in front of the state police car when the authorities arrived and confessed that he had felt like killing because he 'wanted to gather souls'. His defence was temporary insanity due to drug-induced psychosis. Physical substances like drugs can alter our minds and make us do crazy things. Sexton had willingly ingested the drug, but Albert Hoffman, who first synthesized LSD back in the 1950s, also experienced mind-altering trips when he unknowingly absorbed the compound through his fingertips. Simply by touching the drug, his mind was changed. Many toxins can be absorbed by skin contact, and minute quantities of harmful particles can present an invisible threat. Not only do murder and ghosts have to be declared when stigmatized homes are put up for sale, but many US states require houses that have previously contained methamphetamine manufacture labs to be identified and certified clean because of the residual threat of contamination. So when we behave as if houses or clothing could transmit psychosis, we are not being entirely irrational.

However, the fear of contamination does not require something physical. Just the thought of doing something immoral can make us feel physically dirty. It doesn't have to be murder. When adults were asked in a recent study to think about cheating someone, they felt the need to wash their hands afterward.[10] The researchers found that the brain areas that were active when subjects were feeling disgust from physical things like dirt and germs were the same as those that were active when they considered acts of moral disgust. This 'Macbeth effect' reveals that tricks of the mind can be just as powerful as the real thing. So thinking that something could be physically contaminating seems a good reason not to touch it. It is as though we suspect that something like an electric shock could leap from the object. This is why the Fred West cardigan stunt triggers mostly a sense of spiritual, not physical, contamination. You can't wash away such contamination as though it were dirt, but, in a kind of balancing of evil with good, it can be cancelled or 'exorcised' by contact with someone good like Mother Teresa. The Vatican university, the *Regina Apostolorum*, has devised a two-month course on how to carry out an exorcism. From what I understand, these exorcism rites closely follow what was depicted in the classic horror film *The Exorcist*: a combination of prayers, rituals, and commands for the demons to leave the afflicted.[11] The exorcism rite is usually performed in cases of individual possession, and sometimes the sufferer's home is also cleansed with holy water and blessings.

I think that an audience responds as it does to the Fred West cardigan demonstration because most of us would treat the cardigan as if were imbued with evil. In the same way that some of us revere holy sites, priests, and sacred religious relics, we also shun places, people, and objects that are taboo. To do that, however, we have to attribute something more to them than just their physical properties. They must transcend the natural and become supernatural to elicit a disgusted response from us.

WATER COOLER CONVERSATIONS

I have just finished reading *Quirkology* by the British psychologist Richard Wiseman.[12] It's an enjoyable collection of curiosities and factoids about human behaviour, from the search for the world's funniest joke to studies on finding the best opening line when speed-dating. The book is filled with examples harvested from psychological studies, which provide the curious sort of material that people love to discuss in so-called 'water cooler conversations.'

At the end of the book, Wiseman reports the outcome of a series of 'experimental' dinner parties at which people were asked to rate a list of factoids described throughout the book on a scale from 1 ('Whatever') to 5 ('When does it come out in paperback?'). He identifies the top ten factoids that people found most interesting. Here are just the top three most interesting facts. In third place was:

The best way of detecting a lie is to listen rather than look – liars say less, give fewer details, and use the word 'I' less than people telling the truth.

In second place was:

> The difference between a genuine and a fake smile is all in the eyes – in a genuine smile, the skin around the eyes crinkles; in a fake smile it remains much flatter.

Guess what the number-one factoid was?

> People would rather wear a sweater that has been dropped in dog faeces and not washed, than one that has been dry-cleaned but used to belong to a mass murderer.

Now you know why people find this one of the most curious facts about human nature.

WHAT NEXT?

They say that hindsight gives you 20/20 (perfect) vision, and in the cold light of day it is easy to dismiss our reactions to cardigans and pens as irrational when we have all the facts in hand. Whether we knock on wood, wear special tennis shoes, believe we heard a ghost, or avoid objects that may be contaminated with evil, the supersense can be found in many of us.

Some of us are better than others at controlling these thoughts and urges, but we should recognize that they are natural. I think that those with a strong supersense believe that there is more to the human body than simply the physical and that there is a soul or spiritual essence that can leave the body. These are self-confessed supersensers who talk about ghosts and spirits and consult with mediums. However, many of us just feel uncomfortable at the mention of the supernatural. Maybe this is an urge inside most of us that we have to suppress.

I think belief can operate with the same intuitive reasoning that helps us to understand the natural world by letting us make rapid decisions that feel right. The supersense is about these thoughts and behaviours and how they work to bind us together through a belief in invisible forces or essences. They don't all have to be about unearthly experiences. We can use the supersense to connect with each other. Our physicalizing of the spiritual explains our need for contact with those we want to be intimate with, but it also explains how we can castigate others as unclean.

Over the coming chapters, I will tell you unsavoury facts about individuals that will repulse you and make you feel queasy. Such negative reactions reveal that we behave and think as if we can connect with others at a physical level. This in turn produces feelings and emotions that have real consequences for behaviour. In some societies, we may force others to sit on different seats on a bus or keep a certain distance from contact. Segregation and apartheid have been the shameful attempts of some societies to instigate supernatural beliefs about the subjugated members of a group. Such thinking, however, also enables us to see ourselves as connected to our family and ancestors, giving a sense of origin and direction. It explains why heirlooms and birthplace are objects and locations that give us a deeper sense of connection with the past. I think that we do all these strange things because we are social animals bound together by our sense of physical connection. Our thoughts and behaviours extend our individual selves to the group because being a social animal requires reaching out to and connecting with others. Giving gifts, exchanging objects, owning possessions, and making pilgrimages are all examples of our need to make physical connections with others. These connections are not all permanent, but I believe that they are helped by supernatural thinking as we form new bonds and break others. This need is so basic that I am sceptical that rational reasoning could ever get us to abandon it.

Such thinking provides a fertile ground for belief in supernatural phenomena. If you willingly believe in the supernatural, then you are in good company. In a US Gallup poll conducted in June 2005, more than one thousand adults were asked whether they 'believed, were not sure or did not believe' in the ten phenomena listed here.[13] The percentage of believers is reported in parentheses. Take a look at this list. Do you believe any of these phenomena are real?

Extrasensory perception (ESP) (41%)
Haunted houses (37%)
Ghosts (32%)
Telepathy (31%)
Clairvoyance (26%)
Astrology (25%)
Communication with the dead (21%)
Witches (21%)
Reincarnation (20%)
Spiritual possession (9%)

Taken together, most American adults (73 per cent) believed in at least one of the items, while only one quarter (27 per cent) did not believe in any of them. I know what you're thinking: 'Crazy Yanks, they'll believe anything. We're not so gullible here in Europe.' Maybe you, the reader, don't believe in any of these paranormal phenomena, but another Gallup poll conducted on 1,000 Brits at the same time as the US study revealed that we should not be so smug.[14] We are just as likely to believe in haunted houses (40%), astrology (24%), communication with the dead (27%) and the possibility of witches (13%). As a nation, we have no right to point the mocking finger at our American cousins. Yes, they are more religious but we are no more rational. For those of you curious to see how highly you score on paranormal beliefs, there is a self-assessment questionnaire in the Reader's Notes on page 274.

These figures have hardly changed over the last fifteen years and are more or less the same as those produced by the polls conducted in 1990, 1991, 1996, and 2001. Here's my prediction. The figures will be much the same five years from now, and five years after that. I would happily place a large bet on that. I am not a psychic. People are just remarkably consistent and predictable.

To prove this, let me demonstrate my psychic power to read your mind. I bet that you, the reader, also believe in at least one of the items from the list. Go on, be honest. How do I know? First, there is a good chance that you are one of the 73 per cent of the general population who believe. Also, sceptics generally don't bother to read books like this one. In contrast, believers and those who are not so sure want to know whether there is any truth to any of these notions. They understand that their beliefs are considered flaky, and they want to find out whether there is any evidence for things that seem so possible.

There are two reasons to read on. First, supersense is in us all, and I hope to prove that to you over the coming pages. Second, the idea that supernatural beliefs are a product of our own mind design makes it necessary to rethink the origin of beliefs. By examining the evidence mostly from developmental psychology, we can see how such beliefs could emerge in the growing child and how they could continue to influence our thinking as adults even when science tells us to ignore them. This is important, because the development of such notions has relevance to the claim that culture and religions are primarily responsible for creating supernatural belief in the first place.

But don't worry. This book is not meant to make you feel foolish or to encourage you to abandon your supersense. Many facets of our behaviour and beliefs have no rational basis. Think of everything that makes us human, and you soon realize that there is much that calls into question our ability to be rational. Love, jealousy, humour, and obsession, for instance, are present in all of us, and even though we know that our beliefs and actions stemming from them can be unbalanced, we would still not want to lose our capacity to experience them. The same can be said for the supersense. So embrace it, learn where it comes from, and understand why it refuses to go away.

Oh, and if you are a sceptic reading this book, thanks for getting this far.

CHAPTER THREE

WHO CREATED CREATIONISM?

The essence of being human is an uncomfortable duality
of 'rational' technology and 'irrational' belief. We are still a
species in transition.

— DAVID LEWIS–WILLIAMS,
The Mind in the Cave (2004), p. 18

WHO TEACHES US about the 'something there'? When do we start
thinking that there is a hidden but real dimension to reality? Is it
religion, or does religion simply recognize and fulfill that urge in the
human psyche that is so great that we seek out those who explain
why we feel the way we do and then take comfort in their stories,
which make sense of the strange notion that there is something more
to existence? To answer this we have to begin at the beginning.

Two summers ago, my wife Kim arranged for the family to visit
the Niaux cave in the French Pyrenees. It is one of the last Neolithic
caves still open to the public where you can see original prehistoric
cave paintings. Most sites are now closed to protect them from the
destructive moisture and other corrosive properties of human breath.
We booked months in advance, as visits are strictly limited. It may not
be on your list of things to do before you die, but if you want to get
a true measure of the scale of your own life against where humankind

has come from, there can hardly be a more moving experience than marvelling at prehistoric art deep inside the belly of a mountain.

The Niaux system of caves runs over half a mile from the entrance perched high on a Pyrenean cliff face. Outside the temperature was a humid 28 degrees centigrade, but inside it rapidly dropped to a constant 12 degrees. The path was uneven, wet, and slippery, but it was the absolute pitch-blackness that was the most unsettling feature of the caves. The journey varied from claustrophobic passages to wide expanses, created by ancient underground rivers that over the course of millions of years had carved out the inside of the range. Each member of the expedition (I felt like a Jules Verne explorer journeying to the centre of the earth) was given a hand-held flashlight that acted like a light sabre to cut through the ebony shroud. My five-year-old daughter wore those running shoes with lights built into the heels that flashed each time she took a step. She is the fearless type, and she set off with our French guide at the front of the group, picking her way through the tunnel with uncanny ease. The rest of us, unsure of our step, struggled to keep up with the blinking pink flashes that disappeared into the bowels of the earth.

I now understand why people risk their lives exploring underground caverns. The ancient watercourses had sculpted an alien landscape of smooth and bulbous protrusions rising from the floor and dripping from the roof. On the outside, the craggy cliff entrance had been blasted away by modern dynamite, but the inside of the mountain seemed organic and alive. The mica and mineral deposits twinkling in the flashlight triggered childhood memories of Disney grottoes and the seven dwarfs mining for sparkling jewels. Halfway into our descent, we found the hand of man. Mixed in with the graffiti left there by intrepid French youths over the past 350 years was an occasional repeated pattern made up of parallel lines and dots that we were told was much older. Our guide invited us to speculate, but like the experts who carbon-dated the work, we were unable to explain the wood-ash markings put there deliberately for a long-forgotten purpose.[1]

After about an hour, we reached a cathedral-like chamber, the *salon noir*, or black exhibition hall. With our light sabres, we were able to pick out the remarkably well-preserved images of animals and patterns left more than thirteen thousand years ago on the walls of the cavern. This was clearly the focal centre of activity, though no trace of human habitation had ever been found. No bones, no flints, no remnants of someone's lunch. Only the art remained. I tried to imagine the scene illuminated by the flicker of simple lamps made out of animal fat. The place was magical. So often we take for granted our modern lives and all the technologies available to us and easily forget how fast and how far we have travelled. This revelatory experience in the loins of a mountain was a jaw-dropping moment for a twenty-first-century scientist. The people who painted the cave must have thought so too.

David Lewis-Williams studies prehistoric paintings and artefacts. In his book *The Mind in the Cave*, he argues that subterranean art was not for general public viewing.[2] Otherwise, there would be more examples in less remote and more accessible sites. He proposes that the activity in these caves instead reflects early religious attempts to connect symbolically with the earth in its deepest crevices. These places were sacred. The art was deliberately created around the physical properties of each cave. Natural rock patterns and shapes were outlined to form animals in the same way that we see faces in the clouds on a summer's day. This human capacity to see structure and significance in the natural world is not only a talent of the artistic mind but an essential quality for the spiritual one as well. The images came alive through the combination of flickering shadows from tallow lamps and the power of human imagination. Some decorated spaces were only large enough for a solitary individual to squeeze into. The geometric patterns found here may have been the first evidence of the altered states of consciousness that the early shamans are thought to have achieved. Lewis-Williams speculates that the shamans, cocooned in these narrow cervices, sought to document their crossover to the underground world through images and symbols. This may be wild speculation, but what is undisputed is that prehistoric art depicts a mixture of natural and supernatural images. Animals such as horses and bulls, as well as extinct species such as the aurochs and mammoth, are represented, but so are half-human, half-animal creatures.

FIG. 3: 'Lion-Man', a statuette carved of mammoth tusk, dating from around 32,000 years ago, discovered in a cave at Hohlenstein-Stadel, Germany. PHOTO BY THOMAS STEPHAN, © ULMER MUSEUM.

The most stunning example is not a drawing but a statuette from Germany, the Hohlenstein-Stadel 'Lion-Man'. Originally nobody knew what it was. It was shattered into two hundred pieces and mixed among ten thousand bone fragments retrieved from a prehistoric cave in southern Germany just before the outbreak of the Second World War. In 1997 it was carefully reassembled. Who could have predicted how spectacular this find would be? The figure has a human body but a lion's head, stands about 12 inches tall, and is carved from a mammoth tusk. It is not clear whether it is a lion that has taken on human properties or the other way around. Either way, it proves that prehistoric man had imagination and a sense of the unreal. Not only is it one of the most beautiful examples of human art, but it is also one of the earliest. It dates from around thirty-two thousand years ago! Try to get your head around that date for a moment. When it comes to thinking about how long culture and art have been around, we are exceedingly myopic in our outlook.

We may have no written record from this period of humankind, but evidence for supernatural practices can be found in human activity as far back as we can record it. Some of the earliest burials from at least forty-five thousand years ago show signs of ritual. We do not know exactly what motivated prehistoric humans to paint their caves, bury their dead with symbolic objects, or make female ('Venus') figurines with enlarged breasts and stomachs, but these behaviours reflect some of the earliest ceremonial practices in the history of our civilization. Ceremony and ritual have been present from the beginning. There was culture in the caves. Everyday experience must have raised questions in minds sophisticated enough to organize hunts, manufacture jewellery, paint, and communicate. *Where do we go when we dream? What happens when we die?* They must have thought that there was something more to daily existence. Why else spend so much effort celebrating a culture deep in the recesses of a cave if not in the belief that there was something more to reality? From the beginning, humans already had minds prepared for the supernatural.

MODERN MINDS IN THE CAVE

In modern society, we should no longer have a need for shaman to commune with subterranean spirits. Armed with modern science and technology, we can predict and control our lives without the help of trance-induced priests. We can even blast away an entire mountain at the press of a button. We do not have to pray or make sacrifices to control our future. We can measure, test, and document the world. Prehistoric man may have believed in the supernatural, but, then, he did not have the benefit of modern science to explain what he could not understand. Humankind has emerged out of the darkness into a bright, technological, scientific age. By now, we should have abandoned the mind in the cave.

Clearly this has not happened. Over the last four hundred years, we have witnessed an astonishing explosion in our understanding of the universe, something almost like a 'big bang' of scientific understanding. In no other period in human history has humankind made such breathtaking advances in explaining so many facets of the natural world. Wander the corridors of the science departments in any modern large university, and you will find experts in the minuscule details of nature. We have reached out to the farthest galaxies and delved into the subatomic through our science. Science should be the bedrock of our knowledge and wisdom. And yet beliefs in the supernatural – beliefs that are unnatural and unscientific – are still very common.

If science is so successful, why do most people ignore what it has to say about the supernatural? Why doesn't the general public listen to the scientists who say that such belief is unfounded? At this point, I want to draw your attention to the fact that supernatural beliefs generally come in two forms. There are religious supernatural beliefs (God, angels, demons, reincarnation, heaven, hell, and so on) and secular supernatural beliefs (such as telepathy, clairvoyance, and ESP). All religions are based on supernatural beliefs, but not all supernatural beliefs are based on religion. This is an important distinction, since there are some very powerful lobbies and arguments when it comes to the differences between religion, science, and supernaturalism.

As we saw in the last chapter, many Westerners have some form of supernatural belief. The Gallup poll in 2005 revealed that three out of four American adults have at least one secular supernatural belief. Even this figure is an underestimate, for the simple reason that supernatural belief is at the core of every known religion. In the United States, around 90 per cent of the general public is religious, compared to

10 per cent who are atheists.[3] The difference between religious and secular supernatural belief becomes crucial when we consider how we should treat each type. Religious supernatural beliefs are deemed sacrosanct and beyond the realm of scientific analysis. They are miraculous. They transcend the profane and mundane. That's the whole point. Religions must offer unworldly views of reality, not views based in natural laws. Otherwise, they would not be attractive to those people seeking something more from the natural and the ordinary. Religion has to appeal to the *super*natural and the *extra*ordinary. Believers need that spiritual 'X factor' from their religion. In contrast, secular supernatural beliefs are thought to be real phenomena that science has arrogantly failed to acknowledge. All manner of secular supernaturalism has been studied experimentally and, as we shall see, generally rejected by conventional science. Yet in both cases believers have dismissed what science has to say. Why is this?

As we noted earlier, the number one reason people believe in the supernatural is because of their own personal experience. No amount of scientific explanation seems to shake the foundations of such belief. Science seems to make no impact on our supersense. One reason for this is the widening gap between scientists and the general public when it comes to understanding. We are happy to accept the technologies that emerge from science, like the Internet, cell phones, medicines, and so on, but we remain generally ignorant about how science goes about its business. Second, science has a poor public relations image. Ever since scientists were deemed to be tinkering with Mother Nature, they have been held responsible for all sorts of humankind's problems. Today's newspaper headlines about 'Frankenfoods' in reference to genetically modified crops reflect the same deep-seated notion of abomination that was captured so well by Mary Shelley's monster. And what's more, scientists never seem to give a straight answer. They can't agree on the important issues, with experts wheeled in by the media to give opposing opinions that don't seem to provide answers. One day we are told that something is bad and the next, it's good. The public simply don't know what to believe anymore, nor who to trust.[4]

As our planet appears to lurch from one self-inflicted Armageddon catastrophe to the next, from the threat of nuclear holocaust to global warming, many hold the relentless progress of science responsible rather than the technology we have used so conspicuously and greedily. We blame the scientists, not our own human nature. In typically beautiful prose, the psychologist Nick Humphrey summarizes our fear of science:

> Science with its chain-saws and bulldozers of reason, has felled the tropical rainforests of spirituality. It has wreaked ecological destruction on fairyland. It has extinguished the leprechauns, the elves and goblins. It has caused a global change in the weather of imagination. It has made a dustbowl of our Eden, and created an inner drought. And all of this, not to bring greater peace or happiness, but to satisfy people's hunger for the Big Macs of technology.[5]

The nostalgic amongst us look back with rose-tinted glasses and reminisce about a simpler age that seems more wholesome and less threatening than today's uncertain future. We look to ancient cultures for prescientific knowledge, simple living, and spiritual enrichment. We want to get back to nature. We conveniently forget or ignore Thomas Hobbes's callous observation that life in such times was 'poor, nasty, brutish and short.'[6]

Many of us consider modern science a necessary evil. We are happy to reap the benefits of the technology it produces, but deeply suspicious about how it operates. It can be opaque and detached, speaking in a language that make no sense to the rest of society. Any scientist who has briefly stepped into the spotlight has to learn how to explain his or her work in a way that the rest of society can understand. Even scientists from one discipline can be completely unintelligible to those from another. I once took part in the popular BBC Radio 4 science programme, 'Material World' with two astrophysicists.[7] I was talking about the origins of supernatural beliefs while they were arguing about the structure of the universe. I must confess that I felt an acute degree of intellectual inferiority. My contribution seemed trivially simplistic as I struggled to understand their disagreement about whether there are eleven or twelve dimensions to the universe. Phrases such as 'dark matter', 'string theory', and 'multiverses' momentarily triggered faint glimpses of recognition, but since I lacked the necessary skill and experience in mathematics, they might as well have been talking Venusian for all I knew. I expect that's how the public must feel about scientists in general. As he summed up his theory, one of the astrophysicists said that he 'believed' that his theory would be proven right. At last, some common ground for me to enter the discussion. Scientists have beliefs too. They don't always have all the facts. They also have to make leaps of logic in order to put forward a better model to explain the world. The difference between supernatural beliefs and scientific beliefs is that the latter produce testable hypotheses. A good scientist puts forward an idea and, if it fails to stand up to rigorous testing, he or she is obliged to abandon the hypothesis and move on. That's how science progresses – it is always moving forward. In contrast, supernatural believers either do not question their beliefs or ignore the lack of evidence. They do not move forward. In short, the major difference between scientific and supernatural belief is that scientists and believers approach the problem from two completely opposite directions when it comes to weighing up the evidence. Scientists reject beliefs until they are proven beyond reasonable doubt. In contrast, supernaturalists accept beliefs until they are disproven beyond reasonable doubt. The problem is that it is impossible to disprove anything. Logically, you cannot categorically say that something does not exist and will never exist in the future. So you cannot disprove the supernatural. That's why most conventional scientists reject supernatural beliefs as unscientific.

The other important lesson I learned from that day at the radio station is that science may be specialized, but most of us have some opinion on the supernatural. After the broadcast, we all went for a drink at the pub with the production crew. It was not astrophysics that was discussed, but rather supernatural belief. Maybe my fellow scientists were graciously saving me from the embarrassment of not knowing how to discuss the structure of the universe, but they seemed genuinely interested in the public's appetite for the supernatural. During our discussion, it occurred to me that most of us are happy to defer to scientists when it comes to areas of knowledge beyond our own ability. My mathematics is hopelessly mediocre, but I am willing to accept that the astrophysicists know what they are talking about when it concerns the dimensions of the universe. The same must be true for all the other specialized disciplines. However, when it comes to the supernatural, we all have something to say and something we believe. Whether it is our religion or personal conviction that there are supernatural events, science does not have a monopoly on explanations. Also, if the public can see that scientists disagree within their own areas of expertise, then it stands to reason that scientists can't possibly know everything about the supernatural.

And what about belief in general? Belief plays a role in science, religion, and the supernatural. If scientists, priests, and mediums all have beliefs, then who is right? All of them deal with the unobservable, but they rely on different sources of evidence. Science has the scientific method of experiment and observation. The supernatural operates on the basis of personal experience and intuition. Religion is based on culture, testimony, and individual experiences. These descriptions are not perfect, but they capture some of the main differences. Science, religion, and the supernatural are usually treated separately, but we have to consider how they coexist and sometimes overlap in the same mind. I know religious scientists who believe in the supernatural. They bring to mind a Venn diagram showing three circles of beliefs. Some individuals see themselves firmly encamped in one of the circles, but the rest of us are spread out across all three fields. As belief systems, science, religion, and the supernatural are not neatly fenced off from each other but rather blend and blur at the edges, and we cherry-pick from different belief systems when the need suits us. This is important to appreciate as we try to understand the turf wars and tension about belief that have arisen in recent years.

RELIGION AS A VIRUS

In *The God Delusion*, Richard Dawkins denounces all supernaturalism but strategically focuses his attack on the main organized religions.

I decry supernaturalism in all its forms, and the most effective way to proceed will be to concentrate on the form most likely familiar to my readers – the form that impinges most threateningly on all our societies . . . I am attacking God, all gods, anything and everything supernatural, wherever and whenever they have been or will be invented.[8]

Every religion has a supernatural component, but not all supernaturalism is religious. I could be an atheist and still think that I have abilities that go beyond nature but without the need to believe in God. This is important because while all religions come from culture, this is not true for all supernatural beliefs. In making this distinction, we may better understand where supernatural beliefs come from, why they transmit so well, and why they are so difficult to get rid of.

Supernatural beliefs may emerge spontaneously in children as they develop as a natural by-product of their mind design. These beliefs do not need to come from culture. This may also account for why religious beliefs are so successful. Justin Barrett, a religious psychologist, similarly argues that mind design explains belief in God, but I think that such a natural explanation can be extended to all forms of supernaturalism.[9] Religion does not have a monopoly on the miraculous. And if there is a natural origin for all supernatural thinking, then this presents a considerable problem for any attempt to remove supernaturalism, religious or otherwise.

Let's examine the idea that belief is spread by culture alone. We have already seen that we tend to assume that experts know what they are talking about. So it is no surprise that naive children tend to believe what they are told. Maybe our human inclination to believe is something we cannot avoid. It could be immensely adaptive. Such a strategy would increase the learning potential of children by making it unnecessary for them to discover everything by themselves. That's why communicating ideas has been so successful for human civilization. We can learn immense amounts of things about the world without ever having to experience or discover them ourselves. We can learn about people we have never met, places we have never been, and things we have never done and are not likely to do. In fact, we love to learn about things we cannot experience ourselves. However, as Dawkins points out, the same mechanism could be used by the naive child to spread nonsense and lies.

Are children gullible? As every parent knows, the answer is yes, but there are some interesting issues here. Their belief in cultural magical beings such as Santa Claus, the Tooth Fairy, and the Easter Bunny shows that children are open to the possibility of believing the impossible. At the same time, they do appreciate that not all things are possible. Even infants can tell the difference between the possible and the impossible. For example, they recognize a magic trick when they see it. If you hide a toy behind one of two screens and then retrieve the toy from the second screen, as if it somehow moved invisibly from one to the other, six-month-old infants will look longer.[10] Psychologists use such magic tricks to investigate what young children know about the world. If they seem surprised or look longer, then we can say that they noticed something was amiss. Somewhere in their brain, they know something isn't quite right. How then do children come to know what is impossible?

Some knowledge appears to be built into babies by evolution, while other knowledge has to be learned. For example, from the start, babies seem to know the difference between humans and objects and to treat them as very different.[11] Babies interact with people in a totally different way from their interactions with objects. By their first birthday, they have solid objects pretty much figured out, though they are still unsure about nonsolid objects like liquids, sand, and jelly.[12] They can even predict how objects should behave. For example, they know that solid objects cannot float on thin air, and they stare in amazement if shown a conjurer's illusion to create this effect.[13] Are they reasoning about this in a logical way? Are they thinking in terms of why an object cannot float on thin air? When it comes to this type of reasoning, studies have shown that young children reason from experience rather than logic.[14] Children make judgements based on their past experiences. If they have seen something happen, then they know it is possible. However, if they have not seen it happen, they regard it equally as impossible. For example, if told something unlikely – there are people who like to drink onion juice, for instance, or you can find a live alligator under your bed – pre-school children regard these things as being just as impossible as turning apple sauce back into an apple or walking through a brick wall. Only after some years at school can children start to understand that while some things are improbable, they are not necessarily impossible. Children are filtering information through their minds and looking for past experiences to compare it with. This explains why they also deny the possibility that social laws can be broken, for instance, by going barefoot to school or changing the colours of the traffic lights. Since they have never seen any of these events, they regard them as impossible. Also, pre-school children rarely explain why something is impossible. They can't give you a logical argument. Rather, they seem to reason from example. So if you tell them there are things that happen in the world that they can't check out for themselves, they are going to be vulnerable. If they trust you, they will believe you until they have had a chance to check out the truth of what you say.

One analogy often invoked for the spread of beliefs is to compare them to mental viruses or parasites that infect minds. Dan Dennett opens his book *Breaking the Spell* by comparing supernatural beliefs to the tiny parasitic lancet fluke, which colonizes the brains of ants and makes them repeatedly crawl up blades of grass.[15] In doing so, the ant is likely to be eaten by a cow or sheep, thereby fulfilling the next stage of the parasite's reproductive cycle. Dennett is comparing religious ideas to a parasite that makes us spread supernatural beliefs by infecting the child's mind. Strong stuff, and deeply emotive, but Dennett misses an important part of the analogy: both viruses and parasites can infect only hosts that can accommodate them. That is why viruses and parasites cannot infect all species. Viruses can mutate and cross over to different species only after they have changed to fit into the host environment, not the other way around. This minidiversion into virology highlights an important point about indoctrination accounts of belief. Maybe ideas spread not only because children are programmed to believe any idea, but also because they believe ideas that best fit a receptive mind.

Psychologists have long known that we actively have to process ideas in order for them to lodge in our minds. In processing ideas, we compare them to what we know already in order to make sense of them. This can lead to some interesting distortions. Here is a famous example.[16] Consider this description of a young woman:

> Linda is 31 years old, single, outspoken, and very bright. She majored in philosophy. As a student, she was deeply concerned with issues of discrimination and social justice, and also participated in anti-nuclear demonstrations.

Think about who Linda might be for a moment. Imagine a large population of people that includes Linda. Which of these two statements is more likely: 'Linda works in a bank' or 'Linda works in a bank and is a feminist'? Around eight out of ten people consider the second statement to be more likely, but that would be the wrong answer. Consider the problem like a Venn diagram of overlapping groups.

If the number of female bank workers in the world is group A and the number of feminists in the world is group B, you can see that it is impossible to have more female bank workers who are also feminists (A + B) than female bank workers alone. This is because

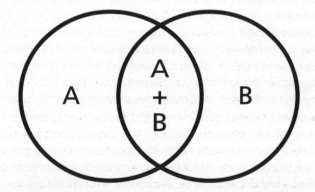

FIG. 4: If the number of female bank workers is A and the number of feminists is B, then there cannot be more feminist, female bank workers (A + B) than female bank workers. AUTHOR'S COLLECTION.

the number of female bank workers who are also feminists will always be a subset of all female bank workers. Nevertheless, the description of Linda seems more typical of a feminist bank worker, and so we say that it is more likely. The Linda problem demonstrates how our minds apply the principle that the more an idea fits with our expectations, the more likely we are to deem it to be true. Our stereotypes of feminists are much stronger than our stereotypes of bank workers, who, let's face it, can seem a nondescript bunch. Because the description of Linda fits our stereotype of feminists, we estimate her to be more likely to be a feminist bank worker, even though there will always be fewer such people in the world compared to all female bank workers.

Why are some ideas more likely? Bankers and feminists are complicated modern concepts that we have learned through culture. Our familiarity with them depends on how often we have encountered these concepts. They do not have any built-in special status. However, other aspects of thinking may be more ingrained in the human mind; traceable to our evolutionary past, they still exert a legacy today. Consider an example that seems more related to irrational thinking and beliefs. Do you have a strong fear of spiders? Does the sight or thought of them make you shiver or feel sick? Do you experience or believe you are faced with great harm when you see one of these creatures? If so, you probably have a phobia.

Phobias are irrational fears and beliefs that are completely out of proportion to the actual source of potential threat. For example, there are no poisonous spiders in the United Kingdom, yet this is one of the most common sources of phobia in that country. Like many wives, Kim makes me remove spiders from the house. I should not complain. We have a friend who also lives in the country but has to pay an exterminator to travel miles to do this job for her when her husband is not around. In 2005 the Zoological Society of London surveyed one thousand adults and found that eight out of ten reported having arachnophobia, the irrational fear of spiders.[17]

It's not just creepy-crawlies. Most of us know someone who suffers from one of the common phobias such as fear of heights, open spaces, snakes, or small dark places. Sufferers can't help themselves. No amount of reassurance or rational explanation can help a truly phobic sufferer. Sometimes phobias become so strong that sufferers cannot stop themselves from taking self-harming actions. For example, obsessive hand-washing is a common symptom of an abnormal fear of contamination. The urge to wash is compelling even though the individual knows that too much washing can be harmful. Such individuals sometimes rub their hands raw until they are bleeding. The 1920s' film mogul and aviator Howard Hughes became famously obsessive about dirt, contamination, and touching other people. He would certainly not wear someone else's cardigan, killer or otherwise.

Where do these beliefs and behaviours come from? Let's consider an explanation based on learning. In the same way that we can acquire superstitious rituals in times of stress, one theory suggests that phobias are caused by a bad experience as a child. In what must be one of the most notorious psychological studies ever conducted, John Watson and Rosalie Raynor presented a nine-month-old baby, 'Little Albert', with a white lab rat.[18] At first the baby showed no fear, but then Watson sneaked up behind the infant and startled him with a loud bang by striking a hammer on a metal bar. Naturally, this startled Little Albert, and he cried. Every time Watson and Raynor presented the rat, they clanged the hammer to frighten the poor child. Very soon, the sight of the rat alone was enough to reduce Little Albert to a shaking bundle of nerves. He had learned to fear the sight of a rat. Little Albert soon became fearful of a number of similar objects that Watson and Raynor presented to him. Not too surprising considering that, whenever these two adults appeared, they seemed hell-bent on making his life a misery. Rabbits, dogs, a sealskin coat, and even a Santa Claus mask soon became sources of sheer terror for the poor child. Only by crawling away could Little Albert get some comfort and relief. He had become phobic to objects that had not previously upset him. These findings supported the theory that adult phobias are due to some bad learning episode as a child.

I know from personal experience that there is some truth to this theory. I used to fish when I was a young boy and didn't particularly like the maggots we used for bait. I remember feeling a bit queasy when I had to pick up their wriggling little bodies to impale them on the hook. It wasn't pleasant, but it was something that I could do. Some years later I would have a terrible encounter with maggots. Like many ten-year-olds, I had taken to searching old derelict houses looking for anything to scavenge. In one house, I remember creeping from one darkened room to another. It had been entirely trashed, as if it had been in an earthquake, and so I had to pick my way through the rubble and household debris. On entering a darkened back room, I heard a faint gurgling, almost buzzing, sound but was unable to see where it came from. I stepped forward onto what I thought was a small furry cushion. In fact, it was the bloated carcass of a dead cat that gave way under the weight of my foot, causing it to pop like a balloon full of rice pudding. Before I realized what had happened, the smell of decay hit my nostrils like a physical punch forcing me to gag and retch. The stench of rotting flesh is universally recognized as one of the most unpleasant on the planet – a response programmed into humans but not carrion beasts or flies. When I lifted my foot into a shaft of light that streamed through a broken window, I stared at the horror of my canvas gym shoe writhing with a mass of maggots. I ran screaming into the daylight and eventually walked home barefoot. From that day on, I have been phobic to maggots. I experience extreme uncontrollable nausea whenever I see them. I particularly hate filmmakers who seem to delight in inserting shots of wriggling maggots into films and documentaries without warning the viewer. As for flies, the creatures that maggots aspire to become, I take great delight in killing them. To hell with karma and Buddhism. If I come back as a fly, I would prefer to be squashed. And do not even think of offering me rice pudding for dessert.

Nobody knows what happened to Little Albert. It is not clear who his parents were and why they would ever have agreed to such an experiment. Watson's study was conducted in 1920; any scientist repeating that experiment today would be fired for unethical conduct. It turns out that Watson was indeed fired, but not for traumatizing Little Albert. In between sessions of terrorizing the baby, he had been conducting an affair with his collaborator. As a married man, his liaison with his graduate student Rosalie was considered too scandalous for the day and so he left academia and went on to earn a fortune in advertising.

The trouble with any learning explanation of phobias is that many patients have never had the sorts of early traumatizing experiences that both Little Albert and I had. For example, it cannot explain people with snake phobias in Ireland or New Zealand, where there are no snakes. Also, if early learning were the only explanation, you would have more cases of car phobias, electric socket phobias, and so on. We are much more likely to have a potentially life-threatening experience with today's technology than with snakes and spiders. It is as if something in our evolutionary past has prepared us to learn these fears. The psychologist Martin Seligman first proposed this preparedness theory of phobias.[19] He argues that humans are genetically wired to fear certain classes of things without the need for a lot of learning. Our species learned to be extra-sensitive to potential threats by natural selection. Maybe our prehistoric ancestors who were especially fearful of snakes and spiders passed that aspect of their personality on to their children through their genes. That would explain why the majority of phobias fit into a few categories that could have been sources or signals of potential danger, such as environments (open spaces, heights, dark places), animals (snakes, spiders), and animals that elicit disgust (rats, mice, maggots). There are few phobias of modern appliances because we simply have not had enough time to evolve wariness to threats like electric sockets.

So some fears seem to take root much more easily than others. Could this be true of other thoughts? Religious beliefs may be indoctrinated by associative learning in the same way phobias are, but like irrational fears, they may also build on our natural inclinations. This is because they fit well with our natural ways of thinking about the world – the mind design we have inherited through our genes. This may partly explain why supernatural beliefs are so easily accepted. They seem to fit with what we think is possible.

This idea of being prepared for the supernatural is one that the anthropologists Pascal Boyer and Scott Atran, who study the similarity of religious beliefs from around the world, have proposed.[20] At first sight, individual religious beliefs seem to be extremely varied, but they all share properties that predict whether they will catch on as ideas. To begin, all religions have a supernatural component – beliefs that violate the natural laws of the world. When Boyer and Atran examined individual supernatural beliefs passed on from one individual to the next through storytelling, they found that these beliefs have a similar structure. First, they were transmitted best when the supernatural aspects were set within a normal mundane context. It was because Jesus turned water into wine at a wedding reception that the miracle was attention-grabbing and is remembered well. His ability to feed a crowd is not particularly surprising until you find out that there were five thousand people and he had only a few fish and loaves. If these supernatural acts had occurred in a much more fantastical context, they would not have had so great an impact. This is called a contrast effect: events are more striking when they suddenly depart from what you expect. That's why horror movies lull you into a false sense of security before the monster jumps out at you. The contrast effect of storytelling has been demonstrated experimentally by showing that the bizarre is best remembered in the context of a normal story line.[21] Completely fanciful stories do not provide such a strong contrast effect, and so they have less impact. Also, the events that violate just one fundamental principle, rather than commit multiple violations, are the most memorable. In other words, the story cannot be too outrageous and fantastical. A statue that speaks is judged to be more likely an example of a 'real' case of the supernatural than one that speaks, bleeds, hovers above the ground, and then vanishes into thin air. The fact that context and credibility are important in the transmission of ideas suggests that people filter stories for plausibility. If this is the case, our intuitive understanding of the world is going to be an important factor in what we believe.

INTUITIVE CREATIONISM

The recent atheist attack on religion has been welcomed by many who are alarmed by the apparent rise and influence of religious fundamentalism in the world. There are a number of reasons for this anti-religion attitude. It is partly a reaction to the perceived rise in the terrorist threat from Islamic fundamentalism around the world that was triggered by 9/11. The reaction is also a response to a corresponding strengthening of Christian fundamentalism and its increasing influence in policy decisions that affect the progress of science and how it is taught in our schools. The battle between science and religion is at its fiercest over the issue of the origins of life on earth, and currently that fight is most bitter in the United States.

The problem is that the majority of US adults believe that a supreme being, namely God, guided the origin and diversity of life on earth. They believe that in the beginning God created earth and all its life forms and that there has been no significant change since that day. This creationist view contrasts with the scientific theory of evolution, which states that life on earth is constantly changing to produce new life forms and that this process continues without purpose, guidance, or design. According to evolution, the diversity of life on the planet we see today is due to gradual changes accumulated over time. The reason this is a problem is that it highlights a paradox of modern America. The United States is one of the most scientifically and technologically advanced nations on the planet. It has produced more Nobel Prize winners than any other country. With the most successful space programme, it has ambitions to colonize neighbouring planets. The United States also has some of the most advanced medical knowledge and practice in the world. Yet less than half of the American population accept a comprehensive scientific theory that explains the origins and diversity of life on earth. When it comes to the general public's acceptance of Darwin's theory of natural selection, the United States is second from the bottom of the list of the top thirty-four industrialized nations. Why is creationism so dominant and natural selection so weak in the United States?

There are two main reasons. First, Christian fundamentalism is politically strong in the United States. In some states, bills have been introduced to allow creationism to be taught as a valid alternative to evolution in the science curriculum. Ever since the famous Scopes monkey trials in 1925, in which a biology teacher was prosecuted for teaching Darwinism, there have been concerted efforts to curb the influence of the teaching of evolution by presenting creationism as a valid alternative. Even though two-thirds of US state science standards recommend the teaching of evolution, fewer than 40 per cent include humans as part of the curriculum. But strong Christian fundamentalism is only one part of the explanation. The other reason creationism is so successful is that there is something about Darwin's theory of natural selection itself that makes it difficult for people to accept. When we see the diversity of life today, it is hard to believe that such complexity could arise spontaneously. Remember: our minds are designed to see order and structure in the world, and everything about life seems to be specially designed, as if by purpose. Darwin's theory explains why this is an illusion. It is beautifully simple, but so alien to the way humans think. To most of us, Darwin's theory of the origin of the species through natural selection is, well, unnatural.

Consider what it has to say. First, we must accept that the world is continually changing. Life on earth has to adapt to those changes in order to survive. Adaptation occurs because each generation of life inherits slight random variations in its genetic makeup from the previous generation, and these variations produce slight differences between individuals. This means that some individuals and not others are better equipped to deal with the pressures of the environment where there is competition to breed. The selection occurs because these individuals are more likely to survive and pass on to their offspring the genes that gave them the advantage. Over time − a lot of time − this gradual process of selection by nature accumulates to produce significant change and diversity.

That's Darwin's theory of evolution in a nutshell. It is a simple, elegant, powerful theory that explains so much about diversity on our planet. But, as Richard Dawkins himself once lamented, it's almost as if the human brain is designed to misunderstand evolution.[22] I think he is right. Evolution is so damned counterintuitive. For example, we can easily see patterns in the diversity of life at any moment in time. However, the same processes that lead us to group animals together also lead us to treat them as separate. As individuals with relatively short life spans, we don't have experience of immense passages of time, and so we cannot observe evolution at work. As laypeople, we don't have the luxury of the historical record to show us how life has changed. All we have as nonscientists are our intuitions about life. And evolution runs counter to those intuitions. How can all living things, from the complexity of humans to the simplicity of bacteria, come from the same original source? How can the complexity of design emerge without a designer? It's precisely because it doesn't fit with our mind design that we find evolution a really hard process to understand.

Also, when people say they are not creationists, are they fully aware of how natural selection works, or are they just rejecting the religious account? Does the rest of the world really understand natural selection any better than the Americans? I am not so sure. In Europe we may readily supply the answer 'evolution' to the question, 'Where did the diversity of life on earth come from?' but, as with many other phenomena, we often say we understand explanations when in fact we don't. This weakness in our ability to be accurate in judging how much we know is called the illusion of explanatory depth.[23] We all typically overestimate how much we understand, and this is especially true of Darwin's theory of natural selection. For example, most people think evolution works by 'survival of the fittest', a term coined not by Darwin but by his contemporary Herbert Spencer.[24] This concept has been misinterpreted to mean that nature selects for those with the most physical strength. This misconception was at the core of Nazi eugenics to kill off individuals who they deemed would weaken the genetic pool. However, this is a gross misunderstanding of the original theory, in which 'fittest' meant how well the individual was matched to his or her environment. It's not always the largest or the strongest individuals who are best matched, because environments are constantly changing, a point Dawkins elegantly explains in his first book, *The Selfish Gene*. If we all evolved into seven-foot-tall, muscular athletes, we would not be very successful in an environment with a limited food supply to feed our massive bodies. This is one consoling fact for those of us lower down the food chain. Eventually those at the top are going to evolve themselves out of existence.

Probably the most difficult aspect of the theory, and the one that smashes headlong into the face of common sense, is the shared ancestry of all life forms. Ever since the Scopes monkey trial, most people have been familiar with the furor over the Darwinists' claim that humans are related to monkeys. But that's nothing compared to the truth about ourselves as revealed by modern genetics. All living things – humans, animals, insects, trees, plants, flowers, fruit, amoebas, and even simple moulds – are genetically related. We know this because science has been able to unravel the building blocks of life and show that all living things share varying degrees of similarity in their DNA structures, the stuff of life. And Darwin's theory of evolution is the only meaningful explanation for this fact. All living things must have evolved from a common ancestor way back in the infancy of life on earth. But, like the argument about whether there are eleven or twelve dimensions to our universe, the science of genetics does not make intuitive sense. From an early age, children treat all manner of living things as fundamentally different in kind. As we shall see, they understand that people are different from pets. Dogs are different from cats. Animals are different from plants. Children are not taught these distinctions. It's a natural way to carve up the living world into all its different forms. Not only that, but children think that all living forms have always existed the way they are today.[25] They are naturally inclined to the creationist's viewpoint.

Like many adults, children cannot conceive of an animal, let alone humans, as a product of constant change. They simply don't have any experience of this, and so they consider it impossible. Of course, we can learn these facts through science education, but they still do not make intuitive sense. That's why we are so fascinated by natural metamorphosis, such as is demonstrated by tadpoles and butterflies. They seem magical because an individual can dramatically change in a lifetime. Actually, metamorphosis in the animal kingdom is not that uncommon. Many species can even change sex, with fish topping the gender-bender list.[26] That might be acceptable for animals, but a transgender human who decides to have a sex change operation is abhorrent to most people – because transgender individuals violate our natural view of humans as being either male or female, a property fixed from birth. In truth, many of our intuitive biological boundaries, such as gender, are more apparent than real. There is much more shared similarity and common origins than we appreciate. And if you don't believe me, ask yourself this: why do men have nipples?[27]

As humans, we do not naturally see ourselves as a product of continual change. Most of us think we are direct descendants in a lineage of ancestors who were also human. That's why we feel a connection with the prehistoric artists of the Niaux caves. However, thirteen thousand years is just a blink of the eye in evolutionary terms. If we go back far enough, we find that life was literally much simpler. I can know this on an intellectual level, but I cannot easily accept that all living organisms have evolved from the same origin. I simply cannot see how I am related to the furry green mould growing on the cheese in my fridge. The full implications of evolution are rarely considered because we cannot conceive what it really means. Our physical resemblance to chimpanzees may make it easier for us to understand that we share around 98 per cent of our genetic makeup. Much harder to accept is that we also share 50 per cent of our genetic makeup with a banana.[28] I may feel that some of my fellow humans have the intelligence of a banana, but to fully accept that all life is related by the same basic genetic building blocks is beyond belief. No matter how simple or complicated an organism can be, all life forms share about one thousand genes. As I write this, I am contemplating the bananas in the fruit bowl in front of me, which for some strange reason suddenly seem less appetizing.

Why do we misunderstand natural selection, and why does creationism do so well in a Christian fundamentalist environment? The answer is that our minds are naturally inclined to a creationist view. After all, creationism was created by the human mind, whereas evolution by natural selection is a fact that was discovered. Without the Book of

Genesis, there would have been some other creation story. The Incas, the Egyptians, and the Aztecs all had exotic creation myths, and that probably goes for all extinct civilizations.[29] Every culture has a creation story because humans are naturally inclined to understand the world in terms of patterns, purpose, and causality. Everything about evolution runs counter to how our natural mind design makes sense of life on an earth made up of different animals and plants. We are not naturally inclined to a theory that is nonpurposeful, nondirected, and yet capable of all the extreme diversity of life forms. To top it all, we are then expected to believe that we are all related to bananas.

Rather, our intuitions from an early age provide a fertile soil for creationism, whether we stumble on it ourselves or are led to it through religious doctrines. These intuitions include:

1. There are no random events or patterns in the world.
2. Things are caused by intention.
3. Complexity cannot happen spontaneously but must be a product of someone's plan to design things for a purpose.
4. All living things are essentially different because of some invisible property inside them.

The developmental psychologist Margaret Evans has studied creationist beliefs in children raised in both fundamentalist and nonfundamentalist homes in the US Midwest.[30] She asked children a series of open-ended questions about the origins of different animals, then coded their responses in terms of whether they were creationist ('God made it'), spontaneous ('It just came out of the ground like that'), or evolutionist ('It came from an earlier different kind of animal'). The youngest children in her group, the five- to seven-year-olds, gave a mix of creationist and spontaneous explanations, depending on their community. As expected, they provided no evolutionary explanations. Also not surprising, those raised in Christian fundamentalist homes were more likely to say that God was responsible, whereas children from nonfundamentalist homes gave an equal mixture of 'God made it' and 'they just appeared' answers.

However, something very strange happens around eight to ten years of age. Irrespective of their home environment, all children of this age gave mostly creationist accounts for life on earth. Something is happening around middle childhood that makes creationism a very appealing explanation to most children. Only at age ten to twelve did children start to show an awareness of evolution and, not surprisingly, this awareness was predominantly shown in the nonfundamentalist households, where families had taken children to natural history museums.

We can know that natural selection is the correct account for the diversity of life on earth, but like the dormant naive reasoning we saw with college students guessing the speed of falling cannonballs, intuitive beliefs can still linger in the educated mind.

RELIGIOUS SCIENTISTS

If God is a delusion and creationism wrong, what can be done to change this state of affairs? It has been suggested that a good grounding in science education can combat the spread of the religion virus. Our best scientists are elected as Fellows of the Royal Society, an august institution that dates back three hundred years to the time of Newton. Around 3 per cent of Royal Society Fellows who responded to a recent survey said they were religious, though I suspect that this figure may be an underestimate, since three-quarters of the Fellows did not respond at all. It may be that religious scientists are aware that their faith beliefs put them in direct contradiction with their science and that they do not want to be 'outed'. There is caution for good reason. When the openly religious member, the Revd Prof. Michael Weiss called, in 2008, for Creationism to be debated as part of the schools' curriculum, he was duly forced to resign as its director of education for bringing the Society into disrepute. It's a similar story in the US. Only 7 per cent of the members of the prestigious US National Academy of Sciences are religious. At first pass, these tiny minorities of 3 to 7 per cent seem to support the idea that scientists are not religious.[31]

The problem with this is that these figures are based on a highly selected group of individuals – the 'A-list' celebrities of the scientific community. The most comprehensive study, conducted in 1969 by the Carnegie Commission, surveyed more than sixty thousand US professors and revealed that around 40 per cent regularly attended church.[32] Of course, society changes over time, and someone who attends church is not necessarily a believer. I had dinner once with Dan Dennett, who surprised me by revealing that he liked going to church. Dennett is famously atheistic and was in the United Kingdom promoting his latest book, which argues that religion is a natural product of mind design. When I heard that he regularly attends church, my jaw dropped into my soup. (I was 'gob-smacked', a quaint British phrase I love, as it captures so well the visual image of one's mouth [gob] when it has been unexpectedly slapped open.) I was aghast. Hold the press. Dennett going to church did not compute until he explained that he enjoyed the choir and the singing. Not all atheists are church-burning militants, and Dennett is still a committed nonbeliever.[33] We were reminded of this recently on his recovery from heart surgery. With typical wit, Dennett thanked those who had prayed for him but wondered whether they had also sacrificed a goat for good measure!

The most recent study, a 2007 survey of 1,646 academics from twenty-one top American universities, reports that only four out of every ten of the physicists, chemists, and biologists interviewed said they did not believe in God.[34] In other words, most of the scientists had some degree of indecision or belief. I find this remarkable, since these academics were from the very 'hard' sciences that demand argument based on objective and reliable evidence. What does this all mean? Basically, that a good science education does not stop you believing in God. Can we really expect the general public to reach the intellectual standards of members of the NAS and Royal Society for them to cease being religious? Science education is essential, and every child can benefit from scientific training, but we must not make the mistake of thinking that science education inoculates the child from religion.

Rather, it appears that culture, not education, is the main factor in the spread of religion. Currently Europe is more secular than the

United States, but that does not mean that Europeans engage in less supernatural thinking than Americans. Atheists can still have supernatural beliefs. A popular poll of one thousand typical UK adults in 2002 revealed that 36 per cent did not believe in God, but nearly twice as many believed that psychics have real powers.[35] As the writer G. K. Chesterton pointed out, when people stop believing in God, they don't believe in nothing, they believe in anything. Even prominent atheists can maintain the possibility of the supernatural. The neuroscientist Sam Harris is a voracious critic of religion.[36] He evokes rational argument to support his attack on faith, and yet, at the end of his book *The End of Faith*, he endorses supernatural aspects of Eastern mysticism and the possibility of the sorts of mental telepathy I address and criticize later. Just because someone rejects conventional religion does not mean that he or she denies all supernaturalism. Some critics quickly denounced Harris's apparent double standards, but I think such criticism is unfair.[37] It is unfair because most of us, including atheist neuroscientists, are naturally inclined to supernatural beliefs.

SUPERS VERSUS BRIGHTS

Dennett argues that we are not all doomed to supernaturalism, since the world can be divided into those with supernatural beliefs ('Supers') and those who reject supernatural explanations of the world ('Brights').[38] I would argue that human nature rarely fits neatly into separate boxes. Such is the case with religion and secular supernatural beliefs. The world does not neatly divide into Brights and Supers on the basis of belief. There is a whole range of beliefs out there. Some beliefs (in heaven, hell, demons, angels, God, and the Devil) are immediately recognizable as the stuff of religious gospel. Other beliefs, such as those surveyed in the Gallup poll cited in the last chapter (precognition, telepathy, clairvoyance) are supernatural notions that contradict our scientific understanding but are not religious. People who say they are atheists can still have some bizarre supernatural beliefs. Most atheists I have met are generally not anti-supernatural so much as anti-religion.

This is a vitally important point that is often overlooked. When I talked in Norwich about humans being wired for a supersense, some thought I only meant religion. Critics pointed out that if we are wired for supernatural beliefs, how can we explain there being so many atheists in countries like Sweden and Finland, where eight out of every ten say that they are not religious? It may be cold up there, but not all Swedes and Finns could have evolved different brains. Or consider a comparison of Ireland and the United Kingdom. Only one out of every twenty is an atheist in Ireland, but skip across the water to the United Kingdom and the number is eight times higher. How could biology explain atheism being prevalent in one country but not in its neighbour?[39]

The answer is that the brain is wired for many things that depend on environment. Just because human behaviour and thinking vary between those raised in different environments does not mean that there is no biology involved. For example, every human infant is wired for language, but the language they end up speaking depends on where they are raised.[40] Infants from anywhere in the world will end up speaking the language to which they are exposed – and with no effort, because their brains are designed to do this.

Or consider an example from vision. Why do all Chinese look alike? Before you start writing to me to complain about my racism, I will add that, of course they don't all look alike, and in fact we also all look alike to them.[41] In an area located just behind your ears is the brain region known as the fusiform gyrus, which is specialized for processing faces.[42] Right from the very start, newborns appear to be wired to seek out faces. With experience, they become expert at recognizing their own mother's face and other members of their group, but they remain less expert at recognizing members of other groups.[43] This research on language and face recognition development tells us there is a biological bias for babies becoming increasingly tuned in to their environment. To borrow an analogy from computing, the infant brain is formatted for certain inputs, and faces and language are just two of them.

TWEEDLEDUM AND TWEEDLEDEE

Could it be that a supersense also results from a biological bias? Maybe culture spreads belief by feeding our bias with ideas, but that does not mean that we inevitably grow up believing. Unlike language and face expertise, which are present in almost every human, belief has much more variation. It depends on the individual as well. For example, I heard a BBC Radio 4 interview with Peter Hitchens and his brother Christopher, who recently published his provocatively entitled criticism of religion, *God Is Not Great*.[44] Both men are intelligent, well-educated journalists. They were raised in the same family, one that taught them to be independent. However, Christopher is an atheist and Peter is a Christian. At the end of a rather surprisingly barbed argument – typical of squabbling brothers, each accused the other of changing the subject – the interviewer interjected and asked how two brothers raised in the same household could be so passionately different in their beliefs. There was a pregnant pause. This simple question had them both lost for words. Eventually, Christopher answered. 'This doesn't help to sell my book!'

The answer to the interviewer's question may be found in a natural experiment that allows investigators to look at the role of biology and environment. When a human egg splits into two after fertilization, the result is identical twins who mostly share the same genes. If these identical twin children are fostered out to different homes, we can estimate the influence of environment and the contribution of genes to their development. It's not a perfect experiment, since most environments are very similar, but it does reveal something fascinating about the power of genes. The research findings are vast, but to sum up the conclusions drawn from identical twin studies, on many psychological measures a comparison of results indicates that it's often like testing the same person twice. Aspects of our personality that we think we have cultivated ourselves are often biologically predictable. This also appears to be true for each twin's inclination toward religion.

Identical twins raised in separate environments share more religious beliefs and behaviour compared to non-identical twins who also live apart. A study by a Minnesota team led by Thomas Bouchard found that the environment is less predictive of religiosity than genetic similarity.[45] Another study from the same group found that once twins leave home, only the identical twins continue to share the same religious beliefs.[46] The geneticist Dean Hamer has even identified a gene, vesicular monoamine transporter 2, or VMAT2, that is linked to the personality traits of spirituality.[47] He found that in a survey of over two hundred people including twins, those who share religiosity also share VMAT2. This gene controls a number of the brain chemicals responsible for controlling moods. Neuroscientists such as Andrew Newberg have even made progress towards identifying the relevant neural circuitry that is activated during religious experiences, again suggesting a brain-based account for the spiritual.[48] So maybe our brains and our own unique mind design determine whether we believe or not. Even if Peter and Christopher Hitchens have shared very similar environments and experiences, they will be pleased to know that they have different brains, which probably explains why their beliefs are so different.

It's early days yet, and it is not clear that reducing the search for belief to the gene level is going to make much sense of a rich and complex human behaviour. However, this research does suggest that the explanation of how belief operates should look at the role of biology working within environments. If the findings from genetic studies hold up, this means that there is something in our genes that contributes to building a brain that is predisposed to belief. If that turns out to be the case, those on both sides of the debate about the true origins of belief are going to be really annoyed, because the suggestion would be that maybe we don't have a choice about whether we believe. In other words, there is no free will in making the decision to believe or not.

Your own individual mind design determines how predisposed to belief you are, a possibility we return to at the end of this book when I discuss mechanisms that control thought processes. However, if there is one thing that both believers and nonbelievers are uncomfortable about it is the prospect that there is a mind design when it comes to choices in life. That's because we like to think that when we make our decisions we are doing so on the basis of objective reason. We like to think that we are weighing up the evidence and making a balanced judgement. In truth, when we make decisions there are all sorts of biases operating that are independent of reason. We don't necessarily have the free will to choose. That's an idea that no one feels happy about. This is because, as the writer Isaac Bashevis Singer observed, 'You must believe in free will; there is no choice.'[49]

EVERYDAY SUPERNATURALISM

Religion is just one form of supernaturalism. You may be a self-avowed, cross-burning, shrine-desecrating, grave-trampling atheist, but I bet that I could quickly uncover some supernatural skeletons in your mental closet. You may also not believe in any of the paranormal phenomena from the ten listed in the Gallup poll from the last chapter, but that list refers only to the ones that are recognized as supernatural. There are many more. For a start, there are the obvious customs like not walking under ladders, throwing salt over your shoulder, crossing your fingers, and so on. These clearly come from superstitious practices passed down through culture. Less obvious are the aspects of normal daily human interaction that arguably reflect beliefs in unseen properties operating in the world. For example, every culture has some form of ritual for greeting that demonstrates the extent to which people are prepared to touch each other physically.

Some cultures are explicit about the supernatural origins of their greeting rituals. The Maori of New Zealand press noses (*Hongi*) to exchange spiritual breath (*ha*), but all contact gestures can be interpreted according to the extent to which there is a perceived exchange of essence. For example, people do strange things in the presence of their idols. Fans go crazy when they get to physically touch their sports heroes or rock stars. Normal, rational people mob the famous simply to make contact. Every presidential candidate has to get used to sore wrists in an effort to satisfy the crowd's desire to shake hands. The need to touch another person is a powerful human urge.

In the same way we are repulsed by psychological contamination from a murderer's cardigan, we are also compelled to engage in acts that address intimate physical contact. Of course, we can always justify them in terms of following traditional customs, but the point is that they originate from supernatural thinking. As a child, did you ever make an oath with a friend in which you both spat on your hands and then slapped them together? You would only do that with someone to make a solemn oath, because touching someone else's spit is so gross. This is because our willingness to make physical contact with others is a reflection of our essentialist beliefs.

Then there are the various beliefs about sacred objects and places. In 2007 John Lennon's piano, the one on which he composed the anthem to humanity 'Imagine', left the United Kingdom to begin a tour of sites around the world. It's not the beautiful white grand piano that we all remember, but a rather plain, brown upright one that you would find in many a school music room. The plan was to take the piano to places of violence and atrocity. It went to the grassy knoll in Dallas where John F. Kennedy was assassinated. It was taken to Memphis where Martin Luther King Jr was shot. It turned up in New Orleans after the devastation of Hurricane Katrina. It appeared at Waco, Okalahoma City, and the Virginia Tech campus – the scenes of so many pointless deaths.

Lennon's piano had become a sacred object to heal the wounds left in communities still coming to terms with grief. Anyone was allowed to touch it. Lori Blanc, a Virginia Tech avian biologist, told me that, even though she is a scientist and not sentimental, she found herself surprisingly drawn to the piano and comforted after playing a tune for a murdered friend. Libra LaGrone, whose home was destroyed by Hurricane Katrina, said, 'It was like sleeping in your grandpa's sweatshirt at night. Familiar, beautiful, and personal.'[50]

All societies have sacred possessions, places, and practices. They become sacred when we attribute special value and powers to them. We believe that they have properties that make them unique and irreplaceable and that no scientific instrument can measure, but most of us believe we can sense them. They are secular supernatural beliefs.

For the sake of argument, let's say that you do not have any of the thoughts I have suggested. However, even the most rational among us can have emotional urges and feelings that run counter to reason. Like Lori Blanc, the scientist who played John Lennon's piano, sometimes we can even surprise ourselves by our own feelings. Cynics too easily dismiss these thoughts and behaviours as simply emotional, as if somehow emotions are less important than reason. However, as my old colleague Dan Gilbert recently pointed out, feelings are the reason humans do anything.[51] Feelings motivate us to go to work, to fall in love, to wonder at the universe, to enjoy life or not. Without feelings, there really would be no point in going on.

Scientists have feelings too. Despite its poor public perception, science can be intensely passionate and emotional. This often comes as a surprise to most nonscientists, but I can tell you that when ideas and reputations are challenged, it can really hurt to be wrong. So I challenge anyone out there to claim they have no emotions. Without emotion, none of us would consider ourselves human. And if you have emotion, I would argue, that emotion cannot be entirely ruled by reason, leaving the door open for the supernatural. We all vary in how much we are influenced by supernatural belief; while many of us can suppress this way of thinking, in the end it is a normal part of the human makeup to reason and behave this way.

Clearly some of us are more prone to this way of thinking than others, but maybe others cannot suppress what is a natural inclination in most of us. We all know what it is to be irrational. Humans are destined to make mistakes of rationality. This irrationality reflects supernatural assumptions that appeal to patterns, forces, and energies categorically denied by science. We don't have our rational radar on all the time. Sometimes our behaviour and decisions are based on inferring the presence of things that science tells us do not exist. That's because the idea of there being something more to reality is such a common ingredient in so much of our human behaviour, irrespective of whether we are religious or not. But I don't want to keep bludgeoning you over the head with the supersense. I hope you will come to the same conclusion. By the end of this book, I want you to reject the idea that you are either a Super or a Bright. Rather, I think it is better to be SuperBright.

WHAT NEXT?

In this chapter, I have dealt with belief in science, religion, and the supernatural. They all depend on thinking about the unobservable, and that requires a mind designed to fill in the missing information. However, by now you should appreciate that this process is not infallible. The same intuitive processes that lead us to reason are the same ones that lead us to be unreasonable. Sometimes we infer the presence of things that do not really exist, and if they did, they would require a complete overhaul of our natural laws. That's what makes them supernatural.

Religion and culture do play a role in the spread of supernaturalism, but I would argue that they simply provide a framework for what comes naturally to us all. We are primed for religious belief because our mind design is biased to supernatural reasoning as a by-product of rational thinking. This subtle distinction between how ideas spread may seem like pedantic hair-splitting but, depending on which is true, there are different implications for what culture can do, if anything, to change supernatural thinking.

Richard Dawkins is right. Religions are spread by culture telling our children stories that have to be believed by faith alone. If we remove the church, religion may be stopped dead in its tracks, but we will still have supernatural thinking. If I am right, it will re-emerge in every newborn child as part of the natural processes of reasoning. It's like the mythical Hydra beast. If you chop off one head it simply grows another. So let's take a look at this monster of a child.

CHAPTER FOUR

BLOOMING, BUZZING BABIES

All human knowledge begins with intuitions, proceeds thence
to concepts, and ends with ideas.
                                    — IMMANUEL KANT,
                        Critique of Pure Reason (1781), p. 569

WHERE DO BELIEFS come from? I'm with the German philosopher
Immanuel Kant on this one. Knowledge generates beliefs, and that
knowledge comes primarily from our intuitive reasoning. Let's examine
the evidence. Most adults are so familiar with storytelling from their
childhood that we assume that what we know and believe comes from
what we were told. However, the picture of the passive child simply
absorbing knowledge and beliefs from others, like some sponge sucking
up ideas, misses an important point. Children come up with their own
ideas long before anyone has told them what to think. Only in the
past fifty years have scientists really begun to appreciate how this thinking
emerges in the growing child. Let me be clear here, because this is the
main argument of the book: *children generate knowledge through their own
intuitive reasoning about the world around them, which leads them to both
natural and supernatural beliefs.* To understand this we have to look at the
beginning again — not the beginning of culture this time, but the

beginning of the developing mind before culture and storytelling have started to play a major role.

The birth of my eldest daughter was a blur for me. As is typical for a first child, the labour lasted a long time, for about twelve hours through the night, and by the time she made her debut the following day around noon, exhaustion, emotion, and sheer anxiety about what was a difficult delivery had ensured that most of my memory of the occasion would be obliterated. Of course, I wasn't the one doing the hard work. My second daughter's arrival was much easier. Well, for me at least. This time I was less anxious, knew what to expect, and frankly was more interested in what various professionals were up to and what the machines were for. Maybe I should have been more attentive to my wife's hardship, but instead I took time to ponder how strange an experience birth must be. I tried to imagine what it must be like to be born – to leave the intimate, warm cocoon of the human womb and enter the sterile, bleached cacophony of a hospital delivery suite, a room flooded with bright light, tubes, cold metal objects, large moving bodies, agitated voices, and machines that go *ping*. What does the newborn make of all this fuss? It's enough to make you want to cry.

In 1890 William James described the newborn's world as a 'blooming, buzzing confusion' of sensations.[1] No organization or knowledge was thought to be present at birth. On entering the world, we were just a bundle of reflexes and dribbles. Reflexes are those behaviours that are automatically triggered. The pupils in your eyes narrow in bright sunshine because of a reflex. When the doctor taps your knee with a hammer and your leg jerks up, that's another. No thinking is required. In fact, you can't stop most reflexes because they are beyond any control or thought.

Babies come packaged with many weird and wonderful reflexes. For example, if you gently stroke the cheek of newborns, a rooting reflex makes them turn their head and mouth to the source of the stroking. They do not know to turn. They are simply wired to do so. There is a sucking reflex when any nipple-sized thing causes babies to pucker up their lips. Clearly these two responses are useful for breast-feeding. There's a stepping reflex where, if you hold the newborn upright with both feet on a surface, it will alternate lifting and placing one leg and

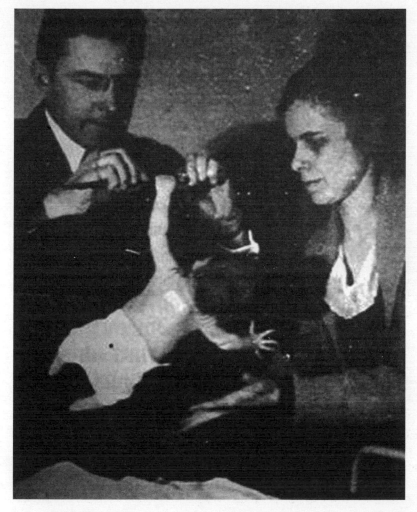

FIG. 5: John Watson demonstrating the strength of the grasp reflex in an infant, dating from around 1919. © THE JOHNS HOPKINS UNIVERSITY, BALTIMORE

then the next in what looks like walking. This astounds parents, because true walking is at least one year off. Then there is the grip reflex.

Their tiny little fingers clamped onto an object placed in their palm are so powerful that you can lift infants off the ground clinging to that object. John Watson did just this, demonstrating something that no caring parent would dream of trying.[2]

In the 'Moro' reflex – sometimes referred to as the startle response – the baby will fling its arms outstretched, as if to hug you if you drop its supported head backwards or make a loud noise. No one is quite sure what that could be useful for. Some of these reflexes clearly support early adaptive functions, whereas others may be a legacy from evolution that we still carry today. Some argue that the Moro reflex was a mechanism by which the prehominid infant grasped onto the furry underbelly of the mother as she fled in dangerous situations.[3] Most modern women with furry underbellies are unlikely to have babies today, but you can still see this primitive response when wild Rhesus monkeys scoop up their babies and scamper when threatened.

As we grow, we lose many of these reflexive behaviours and hold on to others. However, although many of these early infantile reflexes disappear, they are not truly lost, because they can re-emerge in adult patients with head injuries, especially if there is damage to the frontal parts of the brain. For example, in a coma many of the higher control centres of the brain temporarily shut down, allowing behaviours like the grasp reflex to reveal themselves.[4] This is a fascinating feature of our brains, and it may not be limited to simple reflexes. Maybe as we develop we do not entirely abandon all of our initial behaviours and early thoughts. In this way, the brain may be like the hard drive on your computer. Files are never truly deleted, just overwritten but ultimately recoverable.

BRILLIANT BABIES

Apart from reflexes, it was thought that newborns did not have much in the way of what we would call intelligence or knowledge. However, when scientists started to look more closely, they found that newborns are much more aware of their surroundings than simple reflexes would dictate. More striking was the evidence for learning and memory. My own work (the youngest baby I tested was twenty-three minutes old, wrinkled, and covered in afterbirth, but as bright as a button) revealed that newborns can remember and distinguish between different black-and-white-stripe patterns.[5] They also have a preference for faces, as we discuss in the next chapter. This memory for stripes and penchant for faces are something more than simple reflexes could achieve. More amazingly, learning does not begin at birth. For example, if you get pregnant mothers in their third trimester to read aloud passages from Dr Seuss's *The Cat in the Hat*, their unborn babies can hear and remember this experience. When they are born, if you stick a rubber nipple in their mouth to measure their sucking, they will stop when they hear a tape-recording of their own mother reading the same passages. The only way they could have heard this was from inside the womb.[6] Learning clearly takes place before birth. The unborn fetus is listening in on the world and can even remember the theme tune to the TV soap opera that Mum watched during the last months of pregnancy. In one study, the particularly irritating (sorry, memorable) theme tune for the Australian soap *Neighbours* got stuck in babies' heads as much as it did in adults' heads.[7] So be careful what you say. When two pregnant women are talking, there are four individuals listening in on the conversation.

Within a year, most babies can have a conversation with their parents, share a joke, and begin wondering why people do the things they do. They babble, gesture, exchange glances, tease, mimic, and basically become sociable little members of the human race.[8] This transition from the wrinkled newborn in the delivery suite to the socially savvy twelve-month-old is one of the most amazing transformations in life. Something very smart and very fast is happening. We may think computers are smart, but they are nothing in comparison to what a human infant can achieve over twelve months. It is only since engineers started to build computers that we have come to fully appreciate what being smart really is. All the simple things that babies excel at in their first year are some of the hardest problems that engineers have been trying to solve for decades; voice and face recognition, reaching and grasping, walking, reasoning, communication, understanding that others have minds, and even exhibiting humour. All the rudiments of these complex abilities can be found in human infants before their first birthday.

Fueled by the latest research, many parents in the West have come to regard their babies as miniature geniuses, born with unlimited abilities to think and learn. There is now a whole industry of pre-school learning and education that taps into the parental desire to give children the best start in life. By 'the best start in life' what we actually mean is to make sure that our offspring are smarter than the next kid. As they choose among products with names like 'Baby Einstein', 'Baby Bach', 'Baby Da Vinci', 'Baby Van Gogh', 'Baby Newton', and 'Baby Shakespeare', I think that parents' expectations are being somewhat unrealistically raised. In fact, a 2007 study of baby videos and DVDs found that they are associated with impaired language development, a report that infuriated the Walt Disney Company, which owns 'Baby Einstein.'[9]

Parents are easy pickings for those willing to sell them products to enhance their child's future earning potential. We buy black-and-white mobiles to hang over our baby's crib to stimulate the visual areas of the brain (not necessary), chewable toys with bells inside to enhance eye–hand coordination with multisensory input (not necessary), Mozart tapes to improve concentration (myth), flash cards to teach the baby to read (unlikely), and DVDs for the baby to goggle at for hours on end to feed its information-hungry brain (not necessary).[10]

Like gardeners nurturing little plants, we have developed a 'hothouse' mentality to parenting. It's mostly a Western obsession that has more to do with aspirations for our children's success than hard science, but every caring parent is vulnerable. Even my wife, a highly educated medical expert, could not resist the urge to buy the black-and-white mobile.[11] Yes, babies stare at them. They're very noticeable – in the same way that anything black and white is noticeable – but such patterns are not going to accelerate normal growth.

Parents have been cajoled into thinking that natural abilities need a helping hand – or worse, that they can be made better than nature originally intended. Of course, environment is important, but you would have to raise a baby in a dark cardboard box with very little input to produce the sorts of long-term disadvantages that most parents worry about.[12] A normal world with people chatting away, offering attention and affection with food and the occasional toy to play with, is sufficient for nature's programme to unfold. So if you are a first-time parent or grandparent, relax and chill. There is no need for concern when it comes to infant development. It will take care of itself in an ordinary loving household. If a child develops a problem, it's not going to be due to a lack of parental care in a typical setting. It takes severe deprivation to alter the program of normal development. Any concern about understimulation from the environment simply reflects how little we appreciate the complexity of the day-to-day existence that we take for granted.

The image of the brilliant Einsteinian baby was shattered by the following shocking report published in 1997:

Study Reveals: Babies Are Stupid

LOS ANGELES – A surprising new study released Monday by UCLA's Institute For Child Development revealed that human babies, long thought by psychologists to be highly inquisitive and adaptable, are actually extraordinarily stupid.

The study, an 18-month battery of intelligence tests administered to over 3,500 babies, concluded categorically that babies are 'so stupid, it's not even funny.'

According to Institute president Molly Bentley, in an effort to determine infant survival instincts when attacked, the babies were prodded in an aggressive manner with a broken broom handle. Over 90 per cent of them, when poked, failed to make even rudimentary attempts to defend themselves. The remaining 10 per cent responded by vacating their bowels.

'It is unlikely that the presence of the babies' fecal matter, however foul-smelling, would have a measurable defensive effect against an attacker in a real-world situation', Bentley said.[13]

The report went on to reveal that in comparison to dogs, chickens, and even worms, babies also performed the least adaptively when left on a mound of dirt in a torrential downpour. While the other creatures sought cover, the babies just lay there gurgling.

When I last checked, there was no UCLA Institute for Child Development, and I doubt there ever will be following this spoof article written for the satirical publication *The Onion*. These are not the sorts of experiments that scientists conduct on babies, though after reading in the last chapter about John Watson's terrorizing of Little Albert, you might be forgiven for thinking that such experiments are not beyond the realm of possibility. Of course babies cannot defend themselves from attack with a broom handle. They don't need to. That's what parents are for. They are the ones wired to protect their offspring from attack. The article is lampooning the 1993 cover feature for the nowdefunct US *Life* magazine, 'Babies Are Smarter Than You Think.'[14] The cover title went on to proclaim, 'They can add before they can count. They can understand 100 words before they can speak. And, at three months, their powers of memory are far greater than we ever imagined.' Babies may not be able to defend themselves from a broom handle attack, but when it comes to brainpower, they are deceptively smart. Of course, you would be hard-pressed to recognize this. Babies seem so helpless, and, yes, you would think that any creature lying there in the mud and rain is pretty dumb, but you would be wrong. In comparison to a collection of chips, circuits, and transistors, as Marvin Minsky graphically put it, that helpless child is the most amazing meat machine on the planet.[15]

INVISIBLE IDIOTS

It is reported that during the Cold War of the 1960s the American CIA was developing machine speech recognition to translate English into Russian and back again.[16] According to the story, on the debut test-run of one system, the head of operations decided to try out the common phrase 'Out of sight, out of mind.' The computer translated this into Russian, in which it became 'invisible idiot.' 'Out of sight' is indeed 'invisible', and 'out of mind' could mean an idiot. Similarly, 'The spirit is willing, but the flesh is weak', came back as 'The vodka is okay, but the meat is rotten.' These translations make sense literally but bear very little resemblance to the meaning of the phrase in the original language, and they remind us that human understanding requires a conceptual mind, one that can think of ideas and reason over and beyond simple input. As with the colourless green dreams of Noam Chomsky that we encountered earlier, our minds contain information that helps us interpret and make sense.

Even at the basic input stage, our stored knowledge helps us interpret the world. For example, if I were to ask you, 'Do you wreck a nice peach?' I expect you would look at me quizzically. Now, if you ask this question out loud rather than reading it, you hear and understand it as 'Do you recognize speech?', not as an inquiry about whether you are inclined towards destructive acts aimed at pleasurable juicy fruits. You hear one interpretation and not another. This is because destroying a peach is not a common phrase or idea that we entertain. In the same way that we saw the illusory square in chapter 1, our stored knowledge helps us hear and interpret such ambiguous input. We hear one sentence and not another. Where does this knowledge come from? It seems such an obvious answer that knowledge must come from the world of experience. Everything you know must be learned. But is it as simple as that?

Most people are familiar with the blank-slate metaphor that was originally popularized by the British philosopher John Locke in the eighteenth century.[17] The idea is simple enough – children are born without knowledge, and experience shapes them by writing on their minds as though they were blank sheets of paper. Other philosophers, such as Descartes and Kant, pointed out that something has to be built in, otherwise it would be impossible to extract knowledge from a cluttered world of experience.[18] The brain is more like a biological computer that has an operating system we call the mind. That operating system tells us what to pay attention to and how to process information. Without the right operating system, you can't make sense of input – like listening to a foreign language and being unable to understand a word of what is said. Where would you begin? How would you know what you were looking for without some plan? It's like trying to build a house without foundations – you need some embedded structures in the ground to make it stable. The same is true for knowledge. You need rules built in from the start to anchor the information.[19] In other words, you need to be born with some form of mind design. How else would you get beyond James's 'blooming, buzzing confusion'?

AT THE SOUND OF THE DINNER BELL

For many years the importance of mind design was largely ignored in Western psychology. This was partly because in Russia, at the turn of the twentieth century, Ivan Pavlov, working on the physiology of digestion in dogs, stumbled on something that every dog owner knows. Dogs begin to salivate just before you bring them their food. Pavlov called this 'psychic secretion', because it was a reflex behaviour that seemed to be triggered before food was delivered. Dogs are not psychic. They simply learn when dinner is coming by noticing clues such as the sound of the electric can opener in the kitchen just before food arrives. This seems so trivially obvious today, but Pavlov recognized a really important discovery when he saw one – so important that he was awarded a Nobel Prize for it. He realized that animals could be trained to anticipate rewards on the basis of cues. By pairing the sound of a bell with food that naturally causes dogs to slobber, eventually the dogs learned to associate the sound of the bell alone with the impending arrival of dinner. On hearing the bell, the dogs began to drool. It may be my overactive imagination, but I seem to remember a similar response in my old school playground when the bell sounded for lunch. The ringing was enough to make mouths salivate and stomachs rumble.

Pavlov had discovered 'conditioning', a mechanism that would become one of the bedrocks for a whole theory of learning based on association. The idea was that all learning is simple association of events in the environment, like a complex pattern of standing dominoes all stacked up and ready to fall. If you push one over, the others fall in a chain reaction. One event simply triggers the next because of the way the pattern has been formed by association. You do not need to think about a mind making sense of it.

This theory, which provided a way of explaining how babies learn, would dominate Western psychology for the next fifty years. By simply controlling the environment, it was thought that any behaviour could be described and predicted without bothering to know what was going on inside the head. The theory became known as 'behaviourism', and those who followed it treated the mind as a 'black box' that was not only unopened but also ignored. Minds were irrelevant when all behaviours could be described by a set of simple learning rules that created the patterns of mental dominoes.

One of the staunchest early advocates of behaviourism was our old friend John Watson. When he was not tormenting Little Albert, dangling newborns from pencils, or making out with his graduate student, Watson famously boasted:

> Give me a dozen healthy infants, well-formed, and my own specified world to bring them up in and I'll guarantee to take any one at random and train him to become any type of specialist I might select – doctor, lawyer, artist, merchant-chief and, yes, even beggar-man and thief, regardless of his talents, penchants, tendencies, abilities, vocations, and race of his ancestors.[20]

By applying the learning rules of reinforcement and punishment, you can shape patterns of behaviour. If you want to encourage behaviour, give a reward, and an association will be strengthened. If you want to discourage behaviour, give a punishment, and the association will be actively avoided. By linking together chains of behaviour through punishment and reward, it was claimed, the laws of associative learning can shape any complex pattern, be it personality, skills, or even knowledge.

These laws were even believed to explain supernatural thinking. In what was one of the first experiments in irrational behaviour, the Harvard behaviourist B. F. Skinner described in 1948 how he trained birds to act superstitiously.[21] He achieved this with a laboratory box that was wired to give out rewards randomly. For example, if the bird happened to be pecking at some part of the cage when a pellet was delivered, it soon learned to repeat this behaviour. Skinner argued that this simple principle could explain the origins of human superstitious rituals. Like pigeons, tennis players and gamblers seek to reproduce success by repeating behaviours that happened at the time of a reward. Behaviourism explained how something that had long been regarded as a product of feeble thinking could be understood as a consequence of the random reinforcements that the environment occasionally tosses out.

FIG. 6: Deborah Skinner in her father's 'Air-Crib', Skinner's baby crib, in 1945. © *LADIES' HOME JOURNAL*

Skinner would go on to claim that all aspects of child development can be explained by associative learning. He was even accused of taking this too far when he was featured in a *Ladies' Home Journal* article with his infant daughter, Deborah, pictured inside what looked like a giant box similar to the ones Skinner had used to train his animals.

Actually, the box was a special thermostatically controlled crib he had designed for infants so that they did not have to wear baby clothes. In the article, he described the benefits of the 'Air-Crib' as a labour-saving invention that simplified a young mother's life and improved baby welfare. That did not stop the urban myth that circulates today of Skinner raising his own daughter like a laboratory rat.[22] This reputedly led her to grow up psychotic and commit suicide by blowing her brains out in a bowling alley in Billings, Montana, back in the 1970s. Apparently that is a lie. In 2004 Deborah Skinner Buzan wrote an article in the *Guardian* refuting that she had ever been to Billings, Montana.[23]

However, Skinner did go too far with his theories. In the same way that superstitions and rituals emerge, Skinner used behaviourism to explain the uniquely human capacity for language. He proposed that babies acquire a language by a long process of learning words by association, encouraged by their parents to link them together in the appropriate manner. However, when Skinner came to publish these ideas in a book in the 1950s, scientists had already begun to change how they thought about the mind. Behaviourism might have been fine for explaining how the behaviour of pigeons and people can be shaped, but not all human abilities can be taught. This change, known as the 'cognitive revolution', was to become a revolution in thinking.[24]

Skinner was a Harvard heavyweight, but it was a young upstart linguist from down the road at the Massachusetts Institute of Technology who lit the fuse by writing a review of Skinner's book that would go on to become more famous than the book itself. That upstart was none other than Noam Chomsky. Using language development as his test case, Chomsky launched an attack on behaviourism. He pointed out that no association theory of learning could explain how every human child acquires language through learning for the simple reason that the rules that generate and control language are invisible to every natural speaker (unless you are a linguist, of course). Linguists had demonstrated that all the languages of the world share the same deep structures that are hidden from most of us. There is something in our mind design, Chomsky asserted, that we are not privy to but that we can tap when we need to communicate, and this is known as the universal grammar – the invisible laws that govern how language works.

If universal grammar is invisible and most of us are idiots when it comes to linguistics, how can we possibly teach our children by reinforcement and punishment? How can every child acquire language with these hidden rules at roughly the same time, at roughly the same pace, and with little evidence that associative learning plays a role? Something has to be built into the brains of all children that helps them learn language. Chomsky's rapier-like attack dealt a fatal wound to behaviourism from which it would never really recover.

OUT OF SIGHT, OUT OF MIND

The cognitive revolution that took place in the United States did not really happen in Europe, largely because the mind had always been so central in European psychology. In adult psychiatry, Sigmund Freud talked about a fragmented mind in constant conflict with itself. In pattern perception, the German Gestalt School we met earlier, with its meaningful structures and organizations, put the mind at the forefront of human abilities. In the United Kingdom, we had Sir Fredrick Bartlett at Cambridge describing memory as a set of active mental patterns, constantly changing and shifting. But it was in theories of child development that the mind took centre stage as the focus of interest, and no more so than in the theories of the Swiss child psychologist Jean Piaget.

Like Locke, Piaget also had a blank-slate view of newborns, but he thought that they possess learning rules in their tiny minds that enable them to construct knowledge from the apparently simple act of play. Learning and knowledge emerge as babies discover the nature of the world around them in a gradual sequence of revelations. Every simple act of playing with objects – batting them, grasping them, sucking them, pushing them off the high chair – is a mini-scientific experiment for infants, the results of which help form the content of their minds.

Piaget believed that from the start young infants do not understand the world as made up of permanent, real objects but that they treat the world as an extension of their own minds. As though having a bizarre vivid dream. Piaget claimed, infants cannot tell the difference between reality and having a thought. Their world is like the world depicted in the sci-fi blockbuster *The Matrix*, in which evil computers keep the human race in a state of virtual reality by directly feeding experiences into their brains.[25] The computers create the illusion of a normal world. In truth, all the humans are captives, harvested for the energy they produce but completely unaware of their true predicament or surroundings. They are unaware of the external reality that exists outside their minds. In the same way, Piaget's newborns are oblivious to an external reality. They have no concept that the sensations and perceptions they experience in their minds are generated by a real, external world that continues existing even when the baby is asleep. So if some object is really out there, but out of sight, as far as the infant is concerned it does not exist. 'Out of sight, out of mind' became Piaget's signature slogan for this extreme view of the young infant's failure to grasp the permanence of reality. A true understanding of external reality is something babies have to discover for themselves, he claimed, and to do this they need to get interactive.

SEARCHING FOR THE MIND

Somewhere around four to five months, babies get good at reaching and grasping objects.[26] It soon becomes a compulsive behaviour that they just can't stop themselves doing. Any graspable object within reach will do. When my oldest daughter was about this age, I used to carry her on my back in one of those papoose baby holders that leaves their little legs dangling but their arms free to stretch out. When she was not pulling my ears or hair like some demonic monkey on my back, she was always trying to grab anything that came within reach. One day in a supermarket, unbeknownst to me, she reached out and grasped a polythene bag that was hanging from a roll next to the fruit aisle while I was preoccupied with selecting the best apples. I continued walking down the aisle, unaware that I was trailing thirty feet of bags before the smirks of other shoppers alerted me to the growing train of plastic bags behind me.

This fascination with grasping objects is something that Piaget recognized as really important. It means that babies are starting to take an interest in their surroundings. The baby is actively engaging the world. Yet the young infant still does not understand that reality is something different from their mind and independent of their actions. Piaget came to this bizarre conclusion by watching his own young children at play and noticing something peculiar that was to become one of the most famous and studied phenomena in infant psychology. You can repeat this demonstration yourself if you happen to have a six- to eight-month-old infant at hand.[27]

Take one baby and place a graspable object in front of him. So long as he is not already holding something, he will automatically reach out and pick it up, and then jam it in his mouth for taste evaluation. Now unclasp the object and repeat the procedure, only this time quickly cover the object with a cloth and momentarily distract the infant by snapping your fingers. Hey presto! It's gone. It's the easiest magic trick in the world. Most babies will stop and then look around as if the object has disappeared. They do not search underneath the cloth for it. They may pick up the cloth, but rarely as a way of retrieving the object. Because it is out of sight, it is literally out of mind. It no longer exists.

When infants do search under a cloth, a few months later, they still do not understand objects as separate from themselves. For example, if you hide an object under a cushion, a ten-month-old baby will look for it there. But if you then hide the object at a new location in full view of the baby, she will go back and search under the original cushion. The baby believes that his own act of searching will magically re-create the object at the old location. Young children behave as though their minds and actions can control the world. Only through experience do they begin to appreciate the true nature of reality as separate.

MAGICAL BABIES

As it turns out, Piaget was wrong about 'out of sight, out of mind' for babies. We now know that they don't think magically about physical objects. They are not deluded in thinking that their own thoughts make physical things materialize. Babies do know that a real world of objects exists out there. You just have to ask the question in the right way – one that obviously does not require language (because you may be waiting all day for an answer) and does not involve searching for hidden objects. How can this be done? Ironically, the ingenious answer involves a bit of magic.

Everybody likes a good magic trick. Why? Because we don't believe in magic. If we really did think that objects can vanish into thin air, then a conjurer's illusion would be of little surprise to us. Magic tricks work because they violate our beliefs about the world. They cause us to be surprised, to stare in wonder, to look puzzled, applaud, and then want to see it done again. The same is true to some extent with infants. They may not be able to give a round of applause and demand an encore, but they do look longer at the magical outcome of a conjurer's trick. You can measure this simply by the amount of time they spend staring at an impossible outcome in comparison to a possible one.

Over the past twenty years, scientists have used this simple principle to reveal the workings of the baby mind.[28] If babies look longer at a trick, then they must appreciate that some physical law is being broken. Somewhere inside their heads, there is some mental machinery clanking away trying to make sense of an illusion by paying more attention. For example, imagine you are a baby watching a puppet show. On the stage sits a Mickey Mouse doll. A screen comes down to hide the doll, and then a hand comes stage left to deposit another Mickey Mouse doll behind the screen. How many Mickey Mouses (or should that be Mickey

Mice) are there behind the screen? Easy, you say, there are two. But when the screen is raised revealing three, you know something is amiss. The same is true for babies. They look longer at three dolls. They also look longer when only one doll is revealed, but not when there are two. They know one plus one equals two. By five months of age, babies have the basics of mental arithmetic.[29]

Hundreds of experiments have shown that babies can reason about similar unseen events in their head. They can think about hidden objects, where they are, how many there are, and even what they are made of. Where does this knowledge come from? Many such experiments show sophisticated and rapid learning that has led Harvard infant psychologist Liz Spelke to propose that some rules for object knowledge must be built in from birth in the same way the rules for learning language are.[30] Evolution has provided babies with a set of principles to decode the 'blooming, buzzing confusion' that the real world presents to us each time we open our eyes:

Rule 1:   Objects do not go in and out of existence like the Cheshire Cat in *Alice in Wonderland*. Their solidity dictates that they are not phantoms that can move through walls. Likewise, other solid objects cannot move through them.

Rule 2:   Objects are bounded so that they do not break up and then come back together again. This rule helps to distinguish between solid objects and gloop such as applesauce or liquids.

Rule 3:   Objects move on continuous paths so that they cannot teleport from one part of the room to another part without being seen crossing in between.

Rule 4:   Objects generally only move when something else makes them move by force or collision. Otherwise, they are likely to be living things, which, as you will see in the next chapter, come with a whole different set of rules.

How do we know that these rules are operating in babies? For the simple reason that babies look longer when each of them is broken in a bit of stage-show magic. By applying the principles of conjuring and illusion, scientists have been able to show that young infants have knowledge about the physical world that they must be discovering for themselves. And, if they are figuring out the physical world by themselves, then it stands to reason that they must be thinking about other things in the world.

INTUITIVE THEORIES

The things we know best are the things we haven't been taught.
— MARQUIS DE VAUVENARGUES

The magic trick experiments have revolutionized the way we interrogate babies about what they know. If you think about it, all the different things in the world have properties that make them what they are. Inanimate objects have inanimate object properties. Living things have living thing properties, and so on. If you can set up a magic show that violates properties of each of these things, then you can see if the baby spots the mistake.

In a game called twenty questions, you have to work out the identity of something that another player is thinking about. It starts off with the question, 'Is it an animal, vegetable, or mineral?' From there the player has to phrase each question to require a yes or no response. 'Is it bigger than a bread box?' 'Does it come in more than one colour?' If you can guess the identity within twenty questions, you win. An electronic hand-held version called '20Q' won the 2006 Toy of the Year Award from the American toy industry association. It's remarkable. It can almost always figure out whatever obscure object you might have in mind. People find this amazing, but, there again, people overestimate how many different objects they think they know. The reason twenty questions starts with animal, vegetable, or mineral is that this division describes most of the different kinds of things there are in the natural world.

Babies also chop the natural world up into groups of different kinds of things. Not unlike twenty questions, they first decide whether something is an object, a living thing, or a living thing that possesses a mind. From very early on, children reason about the nature of inanimate objects as being different from living things that can move on their own and are alive.[31] They also start to see living things as motivated by goals and intentions.[32] In other words, they are beginning to think about the notion of what it means to have a mind. Well before young children have been taught anything at school, they are already reasoning about the physical world, the living world, and the psychological one. They are in effect little physicists, little biologists, and little psychologists.[33]

However, the knowledge they have in each of these areas is more than just a list of facts. Their knowledge of the world is theorylike. What this means is that when babies encounter a new problem, they try to make sense of it in terms of what they already know. This is what theories do. They give us a framework in which to make sense of something. More importantly, theories allow children to make predictions in a new situation. For example, having established that a spoon pushed off the edge of a high-chair tray falls down, the baby will theorize that other solid objects should do the same and will happily explore this by dropping everything over the edge. The baby is beginning to understand the effects of gravity.

Babies also reason about people. Having witnessed that Mum will pick up the spoon and replace it on the table, they theorize that adults are predictable whereas the family hamster is not. They are beginning to understand that actions differ between living things and to appreciate goals and intentions as mental states. From the moment babies start to pay attention and anticipate events in the world, they are forming theories about how the world works. No one has to teach them about gravity or the mind. They are figuring these out for themselves. It is not even clear that they are fully aware of exactly what they are figuring out, but their thinking is not haphazard. These organized ways of thinking are the intuitive theories that all infants develop.[34]

Most people are familiar with the word 'theory' in the context of science, such as Einstein's theory of relativity or Wegener's plate tectonic theory of continental drift. These are formal scientific theories that have been worked out, discussed, written about, tested, and argued over by hundreds of educated adults. By contrast, children's intuitive theories are spontaneous and naive. However, children do share one interesting property with scientists. Both children and scientists are stubborn when it comes to changing their minds.

CAUGHT IN THE GRIP OF A THEORY

Academics love witty titles for their scientific papers. It not only livens up what could be a very dry piece of writing, but it demonstrates that even scientists can have a sense of humour. In a paper entitled 'If You Want to Get Ahead, Get a Theory', Annette Karmiloff-Smith and Barbel Inhelder describe how children appear to reason in a theory-like way when trying to solve everyday physics problems.[35] The pun is on getting 'a head', which of course can mean either get an advantage or the bony box that houses our brain. However, the paper also makes a very serious point about the role of intuitive theories in intellectual development.

In their study, four-, six-, and eight-year-olds were given wooden rods of different lengths to balance. Imagine trying to balance a ruler on a pencil. How would you go about it? I bet that you would estimate where the middle of the ruler is and balance it on the pencil at this point, which would be the correct solution. The children also balanced the rods in the middle. However, when given rods that were secretly weighted at one end so that they could not balance in the middle, something interesting happened. Initially, all of the children tried to balance these in the middle, but of course they failed. The eldest children looked confused at first, but then realized something was not quite right. They then shifted the rod until they found the point of balance. The youngest children did not seem surprised by the weighted rods and again found the point of balance by moving the rods until they balanced. In contrast, the six-year-olds failed miserably at the task.

Over and over again, the six-year-olds placed the rod in the middle, and every time the rod tipped over. They were so sure that the rods must balance in the middle that they persisted with the strategy until they eventually got frustrated, threw the rods down, and stormed off say-ing that the task was impossible. They were so convinced by the theory that things balance in the middle that they were unable to see that there might be exceptions. This was their theory of balance and, like stubborn adults who refuse to abandon ideas when they are proven wrong, they were unable to be flexible in their behaviour.

Unlike the six-year-olds, the younger children did not have any theory or expectations. They just approached and solved the problem through trial and error. The older children had a theory and also predicted the rods should balance in the middle. However, on discovering this was not so, they had the mental flexibility to realize that sometimes there are exceptions in life. The inflexible six-year-olds were caught in the grip of a theory.

Ten years ago, I discovered a similar phenomenon.[36] Imagine a flexible tube like the one on a vacuum cleaner. Now imagine the tube connected from a chimney to a box below. If I dropped a ball down the chimney, you would know to search for it in the box. You would predict that the ball would fall down the tube into the box. Now imagine that I put a bend in the tube so that the box is not directly below the chimney anymore. If I drop a ball down the chimney, where would you look for it now? In the box of course, because the box is connected to the chimney. What could be easier?

Remarkably, this is something that pre-school children find difficult. They search for the ball directly below. They will search underneath over and over again, even though you can show them each time that it is in the box connected to the chimney by the tube. What's going on?

This weird 'gravity error' reveals some interesting things about the minds of young children. The first is that they reason in a theory-like way. They try to apply the knowledge they already possess to make sense of and predict what might happen next. Just like reticent old scientists, they don't want to believe the evidence when it conflicts with what they expected. All that practice with pushing things off the high chair as an infant has led them to a theory that all objects fall

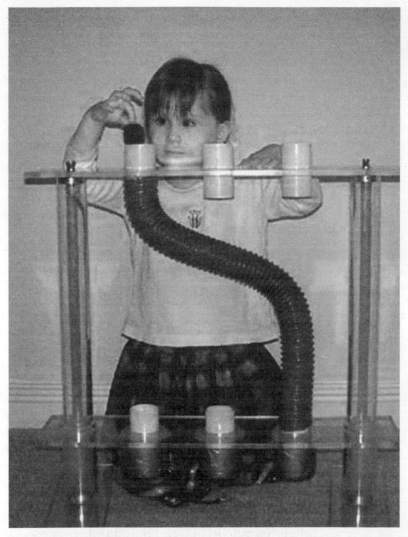

FIG. 7: The tubes apparatus. Children typically search directly below.
AUTHOR'S COLLECTION.

straight down. But when objects don't behave as expected, young children persist with the theory and think something is wrong with the setup. This is because they have trouble ignoring intuitive beliefs.

Humans share the gravity error with chimpanzees, monkeys, and dogs, which have all been tested on tubes.[37] Only dogs seem to learn the correct solution relatively quickly. Are they smarter than young children and primates? Probably not. I think they are more flexible on this task because they don't hold such a strong belief in falling objects in the first place. They are like the four-year-olds on the task of balancing the ruler – not particularly committed to one solution over another.

Eventually children can learn to ignore the gravity error, but even adults can trip up on it. This brings us back to one of the central points in this book. Early ideas may never be truly abandoned. Consider another example from the world of falling objects. What happens to a cannonball fired from the edge of a cliff? Visualize it for a moment. What path would it take? Most pre-school children think that, just like Wile E. Coyote in the *Road Runner* cartoons, the cannonball would travel forward until it loses momentum and then drop straight down.[38] Such beliefs can still operate in adults. If you ask adults what path a bomb takes when dropped from a plane, most of them think that it falls straight down, and they behave accordingly.[39] In games where adults have to release a tennis ball to fall into a cup as they walk by, they typically overshoot because they try to release the ball when directly above the cup.[40] In both examples, the motion is actually a curve, but our childish, naive straight-down theory still operates. These examples show that intuitive theories are not always abandoned when we become adults. If such naive physical reasoning reveals that childish beliefs lurk in adults, what happens if those beliefs are supernatural?

CHILDREN AS INTUITIVE MAGICIANS

When does supernatural thinking first appear? So far in this chapter I have been describing how infants understand the natural world. This process begins long before education has any role to play. Children chop the world of experience up into different categories of things and events. To make sense of it all, they generate naive theories that explain the physical world, the living world, and eventually the psychological world of other people. While children's naive theories are often correct, they can be wrong because the causes and mechanisms they are trying to reason about are invisible. For example, no one can see gravity, but you can assume that something makes objects fall straight down if they are released. Or consider an example from biology. We can easily recognize when something is alive. You can tell by how it looks and more importantly by how it moves, but you can't actually see life in something. All you can do is infer it, and sometimes you will be wrong. Sometimes things do not always fall straight down. Sometimes living things do not move, and sometimes moving things are not alive. When we misapply the property of one natural kind to another, we are thinking unnaturally. If we continue to believe it is true, then our thinking has become supernatural. This is where I think our supersense comes from. Let me unpack this important idea for you further.

Children naturally categorize the world into different kinds of things. If the child is not certain about where to draw the boundaries or misattributes properties from one area to another, the child is going to be thinking supernaturally. For example, if a child thinks that a toy (physical property) can come alive at night (biological property) and has feelings (psychological property), this would represent a violation of the natural order of things. If the child thinks that thoughts can transfer between minds, he is misunderstanding what a thought is and where it comes from. Children who mix up the properties of their naive categories are thinking supernaturally. Inanimate objects that come alive and have feelings are magical. A thought transferring between minds is otherwise known as telepathy.

In hundreds of interviews with children between the ages of four and twelve years, Piaget asked them to explain the workings of the world.[41] He asked them about natural phenomena such as the sun, clouds, rivers, trees, and animals. Where do they come from? Do they have minds? And so forth. What he discovered was recurrent supernatural beliefs, especially in the youngest children. They thought that the sun follows them around and can think. That's why children paint smiley faces on suns. It's much more reassuring to think of it as a friendly being who makes summer days pleasant and people smile than as an inanimate ball of nuclear energy that would frazzle us if it were not for the earth's protective ozone layer. The children Piaget studied believed that trees have minds and can feel. In short, they thought the inanimate world is alive, something Piaget called 'animism'. Animism means attributing a soul (Latin, *anima*) to an entity, and it can be found in many religions as well as in secular supernaturalism. Where do children get these ideas? No one tells them to think like this. It's just the way the child makes sense of the world.

One reason children make these sorts of mistakes is that they make sense of everything from their own perspective. Piaget recognized that young children are so caught up in their own worldview that they interpret everything in the world according to the way it relates to them. Piaget called this 'egocentrism' to reflect this self-obsessed perspective. The sun does appear to follow you around, as it is always there when you look over your shoulder.

Children also attribute purpose to everything in the world by assuming things were made for a reason. The sun was made for me. This is not surprising considering that modern children are immersed in a world of artefacts that have been designed and made for a reason. Young children do not readily make the distinction between things that have been created for a purpose and those that just happen to be useful for a purpose. For example, if I can use a stick for prodding, I may be inclined to see sticks as having a purpose. In other words, sticks exist as something for me to use.

This way of thinking leads the child to what has been called 'promiscuous teleology'.[42] Teleology means thinking in terms of function – what something has been designed for. This way of thinking is promiscuous because the child over-applies the belief of purpose and function to everything. For example, there are 101 ways to travel down a hillside, including walking, skipping, running, roller-blading, skateboarding, sledding, skiing, trail-biking, Zorb balling, and so on. But no adult would make the mistake of saying that the hill exists because of any of these different activities. Children, on the other hand, say that hillsides are for rolling down and so on.

Most seven-year-olds explain the natural world in terms of purpose. As we saw in the last paragraph, promiscuous teleology may incline the child to view the world as existing for a purpose. That's why the creationist view of existence is so intuitively appealing.[43] Most religions offer a story of origins and purpose, which is why creationism fits so well with what seems natural at seven years of age. Maybe that's the origin of the Jesuit saying 'Give me a child until he is seven and I will give you the man.'

Children are also prone to 'anthropomorphism', which means that they think about nonhuman things as if they were human. It's easy to see this happen with pets and dolls, which children are encouraged to treat as human. However, children might also think that a burning chair feels pain or that a bicycle aches after being kicked. They imagine how they would feel if they were burned or kicked and, because of their egocentrism, they misapply this view to everything, including inanimate objects.[44]

Even adults easily slip into this way of thinking. Have you ever lost your temper at an object? Usually it's one that has let you down at a critical moment. The car that dies on the way to an important meeting or, more often in my case, the computer that crashes when you have not backed up your work. Anthropomorphism explains why you talk nicely, beg, and then threaten machines when they act up. It's just the natural way to interact with objects that seem purposeful. We know that talking to objects has absolutely no effect, but we still do it.

So the origins of supernatural beliefs are within every developing child. All of these ideas are not new. The philosopher David Hume wrote about mind design and supernatural beliefs and identified the same aspects of mind design more than two hundred years ago. Hume recognized the same childlike reasoning in adults when trying to make sense of the world. Adults too see a world of things that seem alive with human qualities.

There is an universal tendency among mankind to conceive all beings like themselves, and to transfer to every object, those qualities, with which they are familiarly acquainted, and of which they are intimately conscious. We find human faces in the moon, armies in the clouds; and by a natural propensity, if not corrected by experience and reflection, ascribe malice and good-will to everything, that hurts or pleases us. Hence . . . trees, mountains and streams are personified, and the inanimate parts of nature acquire sentiment and passion.[45]

From this perspective, we can see how an egocentric, category-confused child is going to hold beliefs that are the origins of adult supernaturalism. To begin, children have difficulty distinguishing between their own thoughts and those of others. A child who has an idea thinks that others also share the same idea. Such a notion would be consistent with telepathy and other aspects of mind-melding. Also, children may believe they can affect reality by thinking, which is the basis for psychokinesis: the manipulation of physical objects by thought alone. Children report that certain rituals, such as counting to ten, can influence future outcomes, which is equivalent to spells and superstitions. They also believe that certain objects have special powers and energies. This is sympathetic magical thinking that links objects by invisible connections. To top it all, children see life forces everywhere. Anyone holding such misconceptions could easily succumb to a supersense. This is why I think that adult supernaturalism is the residue of childhood misconceptions that have not been truly disposed of.

DO CHILDREN REALLY BELIEVE?

Do children realize that their misconceptions are supernatural? Young children do not use or understand the word 'supernatural.' Rather, when faced with something inexplicable, they are more likely to say that it is 'magic.' What do they mean by this? The word has lost its sinister connotation and is now used in everyday language. From 'magic markers' to 'Magic Johnson', the term is synonymous with anything special. In recent years, developmental psychologists have begun to question whether children really do believe in magic. After all, they don't try to conjure up cookies when they are hungry, and they know their imaginary friends are just make-believe. In one study, pre-school children were asked to imagine that an empty box contained a pencil.[46] They could do this easily, but they did not actually believe that there was a pencil inside. When another adult entered the room asking to borrow a pencil, the children did not make the mistake of offering them the one that they had imagined in the box.

If adults talk about magic, maybe children are just playing along when asked to imagine magical things.[47] After all, what else would we expect them to say if we tell them that we have hired a magician for their magical birthday party who is going to do magic tricks and that all the guests should come dressed as wizards or fairies? Our whole approach to young children is to emphasize magic as part of normal experience. Magic has lost its supernatural meaning.

The Russian psychologist Eugene Subbotsky revealed magical thinking in young children with a simple conjuring trick. He placed a stamp in a box, muttered magical words in his heavily accented Russian, and then opened the box to reveal a stamp cut in half.[48] Young children believed that it was the same stamp and that the Russian spell had cut it in half. Older nine-year-olds and adults always said that the stamp must have been switched, while younger children were more gullible. But are adults so sure about the world? Although they thought the whole charade was a trick, they were unwilling to put their passport or driver's licence in Subbotsky's box. They did not want to risk being wrong.

When the stakes are high, we are less certain of our reason. It would appear that, just like the killer's cardigan, we consider potential costs and benefits when weighing up the possible unknown. This is why rational students are unhappy to sign a piece of paper selling their soul for real money.[49] Only one in five would put their signature to the contract, even though the form clearly stated that it was not legally binding. Rationally, we would expect them to have more courage in their conviction, like the atheist Gareth Malham, who sold his soul on eBay in 2002 to help pay off his £10,000 student debt. There again, his soul only went for a paltry £10, which hardly justified the effort.

Is it really so surprising that young children give magical explanations for conjuring tricks? Maybe they just use magic as the default explanation when they can't figure something out. What we should find more remarkable is that children grow up in a world full of complex technologies and events that they cannot possibly understand and yet do not talk about them as magic. Remote controls operate machines from a distance. People can talk to others via little hand-held boxes, and so on. The modern child is immersed in a world that would amaze and possibly frighten someone from before the scientific revolution. As Arthur C. Clarke pointed out, 'Any sufficiently advanced technology is indistinguishable from magic.'[50] So why do children not call everything magic?

Young children may start off as Piaget described, with all sorts of magical misconceptions, but with experience older children become more savvy. Children appreciate that there are some things they know and others they do not. When they see something that violates what they expect, they are more suspicious. But it is not the supernatural thinking of young children that is so remarkable, but the supernatural beliefs of adults who should know better. With experience and understanding, supernatural thinking should decline in children, but there is a paradoxical increase in supernatural beliefs in some cultures. In societies where belief in the supernatural is the norm, it increasingly plays an explanatory role in adults' reasoning. This is the effect of environment, and this is where religion wields its influence. For example, when the anthropologist Margaret Mead asked Samoan villagers to give an explanation for why a canoe might have broken its mooring in the night, children tended to give physical reasons, whereas adults were more likely to talk about hexes and witchcraft.[51] That's because the adults had become increasingly influenced by the context of culture.

In our culture in the West, most supernatural beliefs, such as those surveyed in the Gallup poll described in chapter 2, are regarded as questionable, even though the majority of people believe at least one. Adults may even deny supernatural beliefs but, as we noted earlier, so long as no one mentions the word 'supernatural', adults are quite happy to entertain notions of hidden patterns, forces, and essences. In the 1980s, researchers interviewing women in Manchester about their supernatural beliefs found that they had to drop the term 'the supernatural', as this was generally met with negative reactions.[52] However, as soon as the term 'the mysterious side of life' was used, the interviewees showed decided interest and were eager to talk. These women, mostly retired, happily went on to recount multiple experiences of ghosts, precognition, and feeling the spirits of the dead. They regarded these experiences not as supernatural but rather as mysterious.

We have become increasingly aware that supernatural thinking is something to be embarrassed about. We may even hide our superstitious behaviour when there are others around. Three out of four adults will avoid walking under a ladder if they think they are unobserved.[53] If they see another adult do it first, they are much more likely to walk under the ladder. If we do not think that we are being watched, we are more likely to act superstitiously. Students were even less likely to cheat when they were told casually that the exam room was said to be haunted.[54]

Children may not offer an imaginary pencil to an adult, but if left alone, they will check a previously empty box after they have been asked to imagine that it contains ice cream.[55] Even though they know it is just a pretend game, they are still not certain that the ice cream did not somehow materialize inside the box. In another study, four- to six-year-olds were told about a magical box that could transform drawings into pictures.[56] All children denied that such a magical transforming box was possible. However, several days later all of the children tried out the magical spell when left alone with the box and were clearly disappointed when they opened it to find the same drawing inside. This suggests that children do have some expectation of what is and is not possible but are open to the testimony of others. Here is where storytelling and the role of culture can influence children who are uncertain.

Children may not conjure up cookies, pencils, and imaginary friends, but that may be because they understand the limits of their own abilities. They may be less sure about the extraordinary power of others or mysterious magical boxes. This is where culture steps in to shape our beliefs. Again, others' testimony becomes important in supporting supernaturalism, and this is particularly powerful on the playground. In one of the largest extensive surveys of beliefs, Peter and Iona Opie studied more than five thousand British children. Among the various playground activities of games and songs were a mixture of supernatural beliefs relating to oaths and superstitions. The Opies noted that children distinguished between superstitions that were 'just for fun' or 'probably silly' and others that were taken as given. Here the Opies noted the presence of the supersense in those practices that were unquestionably accepted:

> Others, again, are practised because it is in the nature of children to be attracted by the mysterious: they appear to have an innate awareness that there is more to the ordering of fate than appears on the surface.[57]

The other remarkable finding was that children mostly shared the beliefs of their friends but as they became adolescents they increasingly took on the beliefs of their family and elders. The fragmented folklore of children gave way to the traditional beliefs of the culture as they became adults. This may partly explain the pattern of emerging religious beliefs that we saw in the last chapter, where seven-year-olds were mostly creationist in their understanding of the origins of life on earth but older children had started to migrate toward formal religious beliefs or scientific accounts, depending on the family environment.

WHAT NEXT?

So far, the proposal on the table is that the origins of supernatural beliefs can be traced to children's misconceptions about nature. However, this picture is missing a very important piece of the puzzle. No man is an island. We are social animals adrift in an ocean of people. Modern humans have the scientific name *Homo sapiens*, or 'thinking hominid', but as Nick Humphrey has pointed out, the label for modern humans is more appropriately *Homo psychologicus*.[58] Most of our brainpower and the skills that separate us from other animals derive from our capacity to be psychological – to assume that others have minds and reason. This is why we are social animals. We have evolved to coexist in groups, to predict others, to communicate, and to share ideas. All of these skills require a mind sophisticated enough to recognize that others have minds too.

Children's misconceptions may be intuitive and not taught, but they feed into a cultural context to become folklore, the paranormal, and religion. We know that social environments are important in providing these frameworks of belief, but they only exist in the first place because of the supersense. As children discover more about the true nature of the world, they increasingly understand that many of their intuitions are wrong and would only be possible if the supernatural were real. But when others share the same sorts of misconceptions, such beliefs become socially acceptable, despite the lack of evidence or what rational science might say.

In the next chapter, I examine how the supernatural becomes increasingly plausible when we enter the social domain. As *Homo psychologicus*, our social nature depends on our ability to be mind-readers. Each of us is capable of understanding and predicting what others will think and do because we have an intuitive theory of mind. We understand that other people have minds that motivate what they do and what they believe. In the same way that we have intuitive theories of the physical world, humans also have an intuitive theory of the mental one. However, unlike the physical world, where science can objectively verify our beliefs, the mental world is still one of the greatest mysteries that we all take for granted on a daily basis. What is the human mind? How does it work? How does something that is not physical control a physical body? We rarely stop to ask these questions because the mind is so common. Our minds are who we are. It's only when we lose them or they become disturbed that we become acutely aware of how mysterious the mind really is. That mystery is fertile ground for a supersense.

CHAPTER FIVE

MIND READING FOR BEGINNERS

ONE OF THE supernatural powers that I have often thought would be handy is the ability to read other people's minds. Imagine what fun you could have knowing what people really thought about each other. You would know who fancied you (if anyone) or which two people were having an illicit affair in the office. It could make you the most insightful judge or considerate seducer. All the secrets that we try to hide from each other would be out in the open. There again, maybe ignorance is bliss and it is better not to know what others think, especially if those thoughts of others are less flattering about ourselves than we would wish.

We can all mind-read to some extent. Not telepathy or Vulcan mind-melding. That's the stuff of fiction. Rather, we instinctively try to figure out what's on each other's minds. Whether it's winning an argument, negotiating a deal, or serving a customer, all of us recruit our mind-reading skills on a daily basis to infer what others are thinking. We consider what their beliefs might be and guess at which emotions they are experiencing. We want to know 'where they're coming from'. In this way, we anticipate and manipulate others through mind-reading even though we never have direct access to their private thoughts or feelings.

Strangers can read each other's minds when no word has been spoken. As we watch people go about their business in public places, we automatically attribute hidden purpose to their movements. They seem to have intentions and goals. We imbue them with rich mental lives. That's because we think they are like us. They too must experience the same anxieties, disappointments, frustrations, elations, and the whole varied tapestry of human concerns that we do. However, our mind-reading is not foolproof. We often misjudge. Nevertheless, it is easier to understand others as beings motivated by minds rather than the unsavoury alternative: mindless beings, sophisticated robots, or well-dressed zombies.

Some of us are better at mind-reading than others. The Cambridge psychologist Simon Baron-Cohen has proposed that women are more accomplished at it than men.[1] Mind-reading – or social empathizing, to be more accurate – is a female skill resulting from a brain designed to be social. Men, on the other hand, are poor empathizers, but really good at cataloguing CD collections. According to the theory, women are good at empathizing, whereas men are better at systemizing. It's a controversial idea and deeply 'un-PC', but it does seem to fit with much common sense.

Our mind-reading is intuitive. No one teaches us, and we start using it before we can even speak. Like language, it's one of the things that make us human. This is because understanding other minds is so critical to the way we get on with each other. *Homo sapiens* may have evolved to think, but most of those thoughts are about other people. In this chapter, we examine the emergence of mind-reading in our first important relationship with our parents, and in particular our mothers. During these formative years, babies and adults engage in increasingly complex social exchanges. Are you hungry? Do you need your nappy changed? What's she doing? What does he mean? Second-guessing each other is the art of mind-reading, and babies become expert over the early years, better than any other animal.[2] They do this by understanding that bodies are motivated by minds. This understanding equips them for the more challenging role of understanding the social world of others outside the family circle. However, in becoming sociable mind-readers, children start to think about how minds are separate from bodies. That thinking prepares the ground for some very strong supernatural beliefs about the body, the mind, and the soul.

LET'S FACE IT

Our mind-reading starts with the face, and reading the eyes in particular. What do supermodels like Naomi Campbell and Kate Moss, Japanese Manga cartoon characters, and babies all have in common? My, what big eyes they have. One of the reasons we find supermodels and Manga characters so cute is that they remind us of babies. This quality is called 'babyness'. It's simply the large size of eyes relative to big heads on small bodies.[3] Biologists noticed that the young offspring of many mammals share this babyness feature. Puppies, bunny rabbits, and Chihuahuas are all good examples of animals that excel at babyness. It is particularly noticeable in apes because of their large heads, which are needed to accommodate their big brains. However, babyness is more than just a quirk of physical dimensions. For example, if you ask children who have not yet reached puberty to rate faces for attractiveness, they prefer adult faces to baby faces.[4] However, when they hit puberty, girls, in contrast to boys, show a marked reversal by preferring babies to adults. In this way, nature is beginning to pull the strings that shape our reproductive behaviour.

Faces are like magnets to babies. They can't keep their eyes off us. If you measure their eye movements to see where they are looking in a busy social scene, they are checking out the faces of the other people in the room. This interest in faces begins at birth.

For example, given the choice, newborn babies will look longer at the pattern on the left compared to the one on the right.[5] The one on the left looks more like a face than the other, which is identical but upside down. The fact that this is found in babies who have had little experience of faces supports the theory that humans are born to attend to anything that looks like a face. Some argue that this reflects an evolutionary adaptation to make sure babies pay attention to their mother's face in much the same way that young baby birds instinctively

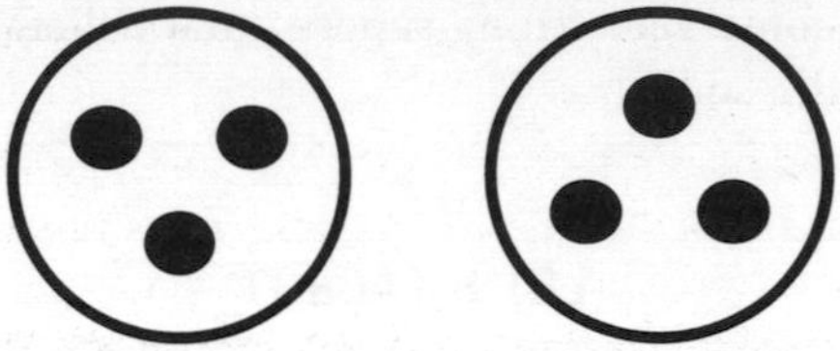

FIG. 8: Newborns stare longer at the face image on the left.
AUTHOR'S COLLECTION.

follow the first moving thing that resembles an adult as soon as they hatch.[6]

So faces are particularly important to humans. We can distinguish and remember thousands of faces, and yet the differences between individual faces can be so small. As we discussed in chapter 3, the fusiform gyrus of the brain (the area just behind your ears) is active whenever you look at faces.[7] However, if you are unfortunate enough to suffer damage to your fusiform gyrus, you can lose the ability to recognize individual faces. The resulting disorder, known as prosopagnosia, can even produce a loss of recognition of one's own face in the mirror.[8]

All this brain machinery dedicated to faces may explain why we are hardwired to see faces when there are none, and often in the most unexpected places. Dr J.R. Harding, a radiologist in Wales, told me about the case of a man who had an undescended right testicle.[9] This condition is common and usually identified during routine screening around the time a boy reaches puberty. Which reminds me of my own experience. I am not sure how the screening is done today, but in my time, before informed consent, most of us prepubescent boys were left completely terrified and perplexed as to why the school nurse asked us to cough as she cradled our scrotums.

When Dr Harding examined the image of the man's descended left testicle, he nearly fell off his seat when he saw what was clearly a face. He wrote the case up and published a medical paper entitled 'A Case of the Haunted Scrotum' for a bit of fun, which became his 'least important but most celebrated contribution to radiology'. In the report, Dr Harding offered an explanation for the absence of the second undescended testicle: 'If you were a right testis, would you want to share the scrotum with that?'

Facelike appearances can readily be found among natural and artificial objects. Boulders, knotted tree trunks, and Volkswagen Beetle cars can

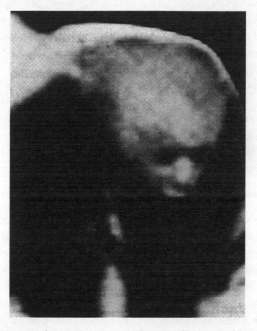

FIG. 9: 'The Haunted Scrotum'. Face image discovered by Dr Harding.
PHOTO © RICHARD HARDING.

all look like they have faces. Because faces are so important, we tend to treat their appearances as auspicious. We think of such appearances as more than just coincidences. In his book *Faces in the Clouds*, Stewart Guthrie argues that our intuitive pattern-processing biases us towards seeing faces, which leads us to assume that hidden agents surround us.[10] Building on David Hume's 'We find human faces in the moon, armies in the clouds' observation that we encountered in the last chapter, Guthrie presents the case that our mind is predisposed to see

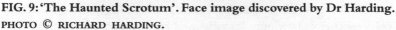

and infer the presence of others, which explains why we are prone to see faces in ambiguous patterns. If you are in the woods and suddenly see what appears to be a face, it is better to assume that it is one rather than ignore it. It could be another person out to get you. Seeing faces leads to inferences of minds. Those minds may have malevolent intentions against us. Why else would they be hiding in the shadows? Such a bias could be just one of the mechanisms that support a sense of supernatural agents in the world. This probably accounts for why face apparitions are often taken as evidence of supernatural activity. For example, the online casino Goldenpalace.com bought a decade-old toasted cheese sandwich said to bear the image of the Virgin Mary for £14,000,[11] and the face of Jesus has appeared on several baby ultrasound scans of pregnant women.

LOVE IS THE DRUG

Faces may be the initial patterns that draw our attention, but it is the emotional experience during intimate moments with those we care about that creates a tangible sense of connectedness. For example, most newborn babies look like grumpy old men with wrinkled skin and bald heads, but to parents these miniature old codgers are beautiful. Mothers can't help falling in love with their babies because nature has slipped them a Mickey Finn cocktail of hormones that forge a passionate bond. Fathers feel it too, but deep down nature really intended this to be a mother–baby thing. It's not as if mothers have a choice. Their bodies are awash with chemical messengers controlling their emotions and behaviour.

One chemical is the oxytocin that surges through the mother's brain around the time of birth to trigger the uterine contractions. It's also active during breast-feeding. Outside of mothering, oxytocin is stimulated by the physical contact of sex. It's no surprise then that research has revealed that oxytocin plays a role in social bonding. Weirdly, we know this because of two species of vole. Prairie voles engage in an intense twenty-four-hour courtship, after which they mate for life. On the other hand, their almost genetically identical cousins, montane voles, are promiscuous and have a preference for one-night stands. They do not pair-bond for life. One explanation is that the reward centre in the brain of prairie voles is sensitive to oxytocin, whereas the same centre in montane voles is not.[12] Oxytocin gives prairie voles that loving feeling because their reward centres are satiated when they mate, but this doesn't happen in montane voles. As Mick Jagger sings, they can't get no satisfaction. When sex scientists blocked reward pathways in the prairie voles, they too became promiscuous with female partners. They did not stick around in the morning or return calls. However, when an injection of a love cocktail including oxytocin was administered to prairie voles, it worked like cupid's arrow, and they bonded again. You could say that those of us who fall deeply in love are behaving just like prairie voles.

When we say the chemistry is just right between two people, there is real alchemy taking place. Sexual attraction and falling in love are experiences enriched with emotions automatically triggered by a cascade of hormones. These are present in the very first social exchanges between babies and mothers but continue to fuel the passion of social intimacy throughout our lives. When this happens, we feel bewitched, enchanted, under a spell, charmed, and generally not in control. Something strange takes hold of us, and rational thinking seems to fly out the window. Breaking down human attraction into chemical neurotransmitters and sensory stimulus patterns may be how science describes the experience, but when Frank Sinatra sang about that old black magic called love, he was describing the supersense that there are mysterious forces at work when people fall in love.

THE RHYTHM OF LIFE

Chemicals and appearance are just two ingredients in the mix of social connectedness. Timing is everything for social relationships too. When two people don't get on, they often say that they just didn't click. We are rhythmic creatures who move in patterns and feel most comfortable with those who move in synchrony with ourselves. Just watch how lovers flirt during a courtship. They exchange glances, utterances, and caresses. If the timing is not right, the relationship is usually doomed.

Movement is also a fundamental way to identify whether something is alive or not. For example, aspects of movement tell us when we are dealing with an animal or an object. Objects move in a rigid fashion, whereas animals have a fluid, groovy motion. The next time you are in a shopping mall, watch how other people move. Smoothly and fluidly, shoppers steer and glide past each other to avoid collisions. Machines couldn't negotiate a busy street full of people. Second, the type of movement is instantly obvious. If you attach luminous reflectors to a person's forehead, elbows, wrists, knees, and ankles, then turn the room lights off, you see nine separate glowing spots in the dark. However, as soon as that person moves, you immediately see him or her as a person.[13] Stop and the person becomes nine stationary points again. That's because our brains are exquisitely tuned in to the smooth movements of living things even when we can't see their bodies. It's so fundamental that when shown these lights point displays, even babies as young as four months see the invisible person.[14]

Like faces, sometimes movement can fool us into thinking that something has a mind. For example, toys that seem to come alive fascinate children. In my day, one of the popular toys was a piece of finely coiled wire called a 'Slinky'. It could appear to walk by stretching and lifting up one end over another down an incline, a bit like an acrobatic caterpillar. The attraction of the Slinky on Christmas Day was the lifelike movement it had as it stepped down the stairs before someone trod on it or twisted the spring and ruined it for good. Toys that appear to be alive are curiosities because they challenge how we think inanimate objects and living things should behave. Many toys today exploit this principle to great effect, but be warned: not all babies enjoy objects that suddenly seem lifelike. This anxiety probably reflects their confusion over the question, 'Is it alive or what?'

Once babies decide that something is alive, they are also inclined to see its movements as purposeful. They are beginning to infer a mind controlling the movements. In one study, twelve-month-olds faced a stuffed toy on a pedestal.[15] It looked like a kind of furry brown Russian hat known as a 'shapka', with two button eyes for a face. Hardly the most convincing example of a living creature. However, unbeknownst to the baby, the shapka was remotely controlled by scientists hidden in another room. The baby watched the shapka. The shapka watched the baby. It was like the standoff in a spaghetti western. After a short uncomfortable silence, the hat suddenly beeped and moved. The baby was surprised and looked towards its mother for some explanation. None was offered. The baby pointed at the shapka and vocalized. The hat responded back with beeping. The scientists controlling the shapka made sure that it responded to every utterance and movement the infant made. Very soon the baby and hat were engaged in a meaningless but richly synchronized social exchange. When the hat swung around as if to look off to the side, the baby followed suit to see where the shapka was looking. The baby was treating the hat as if it had purpose. Simply by interacting in a synchronized way with the baby's own responses, the shapka and baby had become the best of buddies.

Babies respond to such exchanges as if the objects are alive and have purpose. They infer intentions. However, if the shapka had simply moved randomly and had not had a face, this social connection would not have been made, and the babies would not have copied or tried to follow the hat's lead. So movement and faces lead to the inference of intentional purpose. It's such a powerful combination that it is almost impossible to ignore.

HERE'S LOOKING AT YOU, KID

Humans are natural people-watchers, and most of the time we look at faces and eyes. The focus of another person's gaze is a very powerful signal for us to look in the same direction. Magic Johnson was a great basketball player because he used the 'no look' pass: he could pass the ball to a teammate without taking his eye off his opponent.[16] He could control his gaze to hold the other player's attention and not betray with his eyes where he was about to pass. More impressive was his ability to look toward one teammate and then pass to a completely different person, sending the defender on the opposite team in the wrong direction.

Our difficulty in ignoring the gaze of another person shows what an important component of human social interaction it is.[17] They say that the eyes are a window to the soul. I don't know about souls, but eyes are a pretty good indicator of what someone may be thinking. You can observe this yourself the next time you are standing in line at the supermarket checkout. Just watch the rich exchange of glances between people. It's remarkable that we are often so unaware of how important the language of the eyes is. This is one reason why it is so unnerving to have a conversation with someone who is wearing sunglasses and we cannot monitor where they are looking. Police officers wear mirrored sunglasses to intimidate suspects for this very reason.

This sensitivity and need to see another's eyes is present from birth. Newborn babies prefer that we look them in the eye. Even though their vision is so poor that they would qualify for disability allowance,[18] they can still make out the eyes on a face, and they prefer the faces of adults whose gaze is directed towards them.[19] Since they have little experience of people-watching, this strongly suggests that gaze-watching is another process built in at birth. People in love stare at each other, and parents and babies spend long periods engaged in mutual staring. If you look into the eyes of a three-month-old, the baby will smile back at you. Look away and the smiling stops. Look back and the baby smiles again. Mutual gaze turns the social smiling on and off.[20] Not surprisingly, it works in the other direction. If the baby stares, parents smile. They really do have us wrapped around their little fingers.

Gaze is part of a general range of social skills called joint attention.[21] When humans interact socially, they do so by sharing the same focus of interest. Whether it is discussing a topic, watching a basketball game, or admiring a painting, we can join in a combined effort to examine the world. Joint attention is not uniquely human; many animals use it to extend their range of potential interests or threats. Like meerkats, who watch each other for the first sign of danger, animals can gain the benefit of watching others watching the world. However, the jury is still out about whether other animals can infer the mental states that humans appear to infer.[22] Consider this passage from Barbara Smuts's 'What Are Friends For?':

> Alex stared at Thalia until she turned and almost caught him looking at her. He glanced away immediately, and then she stared at him until his head began to turn toward her. She suddenly became engrossed in grooming her toes. But as soon as Alex looked away, her gaze returned to him. They went on like this for more than fifteen minutes, always with split second timing. Finally, Alex managed to catch Thalia looking at him.[23]

Smuts suggests that Alex and Thalia could be two novices at a singles bar. In fact, this description comes from her field notes of two East African baboons beginning a courtship. It could have been lifted straight out of a scene from *Sex in the City*, although I would guess that a woman suddenly grooming her toes in public might be considered a bit of a turn-off in New York's downtown Manhattan. Are animals capable of mind-reading? Certainly they seem to follow gaze, but it is not clear that they can really get to the next stage, which is to think that others have mental states such as beliefs and desires. That is something that seems to be a particulary human quality, and one that infants achieve somewhere between their first year and second year.

THE GOOD SAMARITAN

Being able to understand others as having goals is a powerful mind-reading tool. It allows us to interpret other people's actions as being purposeful and also allows us to anticipate what they might do next. Consider the following sequence of events as if watching a silent movie. Our intrepid climber approaches the steep hill and begins his ascent of the slope. Halfway up the hill, the climber comes to a level where he stops momentarily before resuming his journey. On top of the hill, another person is waiting. Suddenly this other person charges down the hill, blocking our climber's progress and forcing him down the remaining slope with repeated shoving. What's going on here? Is this about a land dispute? Or maybe they are dueling for the hand of the maiden who dwells at the top of the hill? What most people assume is that there is a clash of interest and that the two are not friends. In another version of the movie, instead of hindering our climber's ascent, another individual comes along and helps the climber up the slope. Again, a fertile imagination could construct a feasible explanation. Is he a Good Samaritan who helps climbers up the hill?

Actually, the two events are computer animations used by Yale psychologists to investigate the origins of human morality.[24] The various players in these mini-dramas – the climber, his assailant, and the Good Samaritan – are in fact geometric shapes with eyes simply moving around a computer screen. But when you watch these sequences, you cannot help but see them as purposeful individuals with goals and personality. At work here is the anthropomorphism that we described in the last chapter. Even simple geometric shapes seem alive if they move by themselves, taking paths that seem purposeful. Our anthropomorphism endows the shapes with humanlike qualities of mental states. By hijacking the rules for the movements of living things and applying them to objects, we effectively make them come alive.

Just like you or me, the twelve-month-old babies watching these sequences also judge the nature of each shape as good or bad based on the way it behaves. Long before we have a chance to teach infants about good and bad people, infants are making these judgements by simply watching social interactions. First, the climber is seen as purposeful with the desire to reach the summit. The assailant who forces the climber back down the hill is nasty, whereas the one who helps the climber is nice. We know this because if the helper or hinderer suddenly changes behaviour, babies notice the switch. Babies know something about the nature of the individual players. Not only that, but when later offered a replica toy of the helper or hinderer to play with, almost all babies choose the helper doll. Babies prefer to play with the Good Samaritan.[25]

If, after pushing the climber downhill, the hinderer is painted so that it now looks like the Good Samaritan, babies are not fooled by the change in outward appearances. They know that deep down it is still a nasty piece of work because they are surprised if it suddenly starts helping the climber again. Babies know that appearances may be deceptive and that being bad is a deep personality flaw. As the saying goes, 'A leopard can't change it spots.'

SECRET AGENTS

Whether it's heroic geometric shapes, animated toys, or contingent Russian shapkas, mind design forces us to treat such things as if they have purpose and goals. Our natural tendency to assume that people's behaviours are motivated by minds allows us to predict what they might do next. This is what Dan Dennett calls adopting 'the intentional stance'. When we adopt the intentional stance, we detect others as agents. An agent here is not James Bond, but rather something that acts with purpose. We attribute beliefs and desires to agents, as well as some intelligence to achieve those goals.[26] This could be an adaptive strategy to ensure that we are always on the lookout for potential prey and predators. By adopting the intentional stance, you are giving yourself the best chance in the arms race of existence to find food and avoid being eaten.

However, the trouble with assuming an intentional stance is that it can be wrongly triggered. Things that don't have intentions but seem to – because they either look as if they are alive (movements and faces) or behave as if they are alive (respond contingently) – make us think they are agents. We are inclined to think that they are purposeful and have minds. There's a company in Somerset where I live that makes a vacuum cleaner that has a face painted on the front called a 'Henry'. Actually, it's called the 'Numatic HVR 200–22 Red Henry vacuum cleaner', but people know it as 'Henry' for short. From reading the customer reviews on the Amazon website, where you can buy the vacuum cleaner online, it seems to be a fine little sucker. What is

FIG. 10: The 'Henry' vacuum cleaner. © NUMATIC INTERNATIONAL LTD.

surprising is the way people describe the vacuum cleaner. Henry is not referred to as a machine but rather as a 'he', 'a loyal servant', and so on. As one customer put it: 'We've had our Henry for about 14 years. He cleans the house, the car and DIY dust without a complaint . . . and he's always smiling. How many of your household appliances do you apologize to if you accidentley [*sic*] bang it on a corner as you go around?' Henry clearly triggers a very strong intentional stance in his owners.

I don't think anyone really believes that Henry is alive or has feelings. But the vacuum cleaner does illustrate how easy it is to adopt the intentional stance. This may not be such a bad thing. After all, when we are trying to understand and predict events in the world, adopting an intentional stance gives us a useful way of framing information and doing things. For example, let's say my car breaks down one day. Confronted with this, I have to plan a course of action to fix the problem. What's *troubling* her? Maybe she *wants* a service. The old girl *needs* a face-lift. Dennett gives another good example.[27] Gardeners *trick* their flowers into budding by putting them in the hothouse so that they *think* it is spring. The intentional stance is just a comfortable way of talking about and interacting with the natural and artificial world. But as we saw in Piaget's animism in children, this way of thinking emerges early and may support a supersense that there are secret agents operating throughout the world. It is supernatural because it represents the over-extension of the intentional stance from real agents with minds to objects that cannot have this kind of mental life. Certainly we slip into this supernatural way of thinking remarkably easily. We may laugh it off, but as the saying goes, there is no smoke without fire. It must have some influence on our reasoning, lurking there in the back of our minds. The very same processes that led us as babies to seek out potential agents in the world continue to fool us as adults into thinking that the world is populated with purposeful and willful inanimate objects.

GHOSTS IN THE MEAT MACHINE

Whether we are reading our own mind or inferring the mind of others, we are treating minds as separate from bodies. This idea that the mind exists separately from the body is known as 'dualism'. In his book *Descartes' Baby*, Paul Bloom heralds an impressive avalanche of work to argue that humans are born to be intuitive substance dualists.[28] Substance dualism is the philosophical position that humans are made up of two different types of substances, a physical body and an immaterial soul. Our mind is part of this soul that inhabits our body. The separation of the body and mind – or the 'mind–body problem', as it is known – is one that keeps philosophers and neuroscientists awake at night. Let me explain why.

Each of us experiences our mental life as distinct from our body. We can see how our bodies change over the decades, but we feel that we remain the same person. For example, I think I am still the same man I was in my late teens. I sometimes still behave that way. Our knowledge, experiences, ambitions, priorities, and concerns may change over the years, but our sense of self is constant. This is one of the most frustrating aspects of ageing. Old people do not feel they have aged; only their bodies have. And what's worse is that Western society is increasingly ageist. We treat old people differently and patronize them. But old people feel they are generally no different from when they were young. When we look in the mirror, we can see how the ravages of time and gravity have taken their toll on our bodies, but we still feel we are the same self. We may even change our beliefs and opinions with time, realizing that some punk music was actually pretty awful, but we don't experience a change in the person having those beliefs or opinions. That's because we cannot step outside of our mind to see how it looks from a different perspective. We are our minds.

In addition to the cruel injustice of youthful minds being trapped inside ageing bodies, our daily experience constantly tells us that our minds work independently and in advance of our bodies. Every waking moment, we make decisions that precede our actions. It seems that our bodies are controlled by our thoughts. We feel the authorship of action. We are the ones doing the doing. This is the experience of conscious free will. However, free will – the idea that we can make whatever choices we want, whenever we want – is most likely an illusion. The experience of free will is very real, but the reality of it is very doubtful.

Cognitive scientists (those who study the mechanisms of thinking) believe that we are in fact conscious automata running a complex set of rule-based equations in our heads. We are consciously aware of some of the outputs from these processes. These are our thoughts. We experience the mental processes of weighing up evidence, considering options, and anticipating possible outcomes, but the conclusion that our minds have a free will in making those decisions is not logical.

If you doubt this (and most readers will), then consider this. If we are free to make decisions, at what point are decisions made and who is making them? Who is weighing up the evidence? Where is the 'me' inside my head considering the options and doing 'eeny, meeny, miny, moe?' That would require someone inside our heads, or a ghost inside the machine. But how does the ghost in the machine make decisions? There would have to be someone inside the ghost's head making the choices. So if there is only one ghost, how does it arrive at a decision? Does it look at all the alternatives and then flip a coin? If so, flipping a coin can hardly be free will.

THE NUMSKULLS

My editor tells me that these are really difficult concepts that need explaining, so rather than ghosts inside heads flipping coins, let me tell you about 'The Numskulls'.

When I was a kid growing up in Dundee, Scotland, 'The Numskulls' was the local DC Thompson's comic strip about an army of little people who lived inside of the head of a man called Edd. They were workers controlling his body and brain. And like workers in a factory, sometimes they would screw up. For example, the Numskull controlling the stomach would see that reserves were getting low and send a request for more food. The Numskull responsible for feeding would pull the levers to get Edd eating. Maybe the Numskull in the tummy would fall asleep at his station because of all the food, and Edd would end up stuffing himself until he became sick. An alarm light would go off in the brain department, where the boss Numskull sat at his executive desk reading the incoming messages. Then there would be a frantic race to tell the eating Numskull to stop working. You can see how such a scenario easily generated comic story lines each week as the machine called Edd would encounter different problems arising from his internal workforce. It was one of my favorite comics, even though I did not realize that the creators were actually presenting children with a profound philosophical conundrum about free will.

FIG. 11: 'The Numskulls' from my childhood © D.C. THOMSON & CO, LTD, **Dundee.**

The Numskulls show that decision-making is a deep problem. How are decisions arrived at? If a choice has to be made, how does that happen? We intuitively think that we make the decisions. We make up our minds. But how? Is there a Numskull boss inside my head? And if so, who is inside his head, and so on? Like an endless series of Russian dolls, one inside another, an infinite number of Numskulls becomes an absurd concept.

To cap it all, the experience of conscious decisions preceding events may also be an illusion. If I ask you to move your finger whenever you feel like it, you can sit there and then eventually decide to raise your digit. That's what conscious free will feels like. But we know from measuring your brain activity while you're sitting there waiting to decide that the point when you thought you had reached a decision to move your finger actually occurred after your brain had already begun to take action.[29] In other words, the point in time when we think we have made a choice occurs after the event. It's like putting the action cart before the conscious horse. The mental experience of conscious free will may simply justify what our brains have already decided to implement. In describing this type of after-the-fact decision-making, Steven Pinker says, 'The conscious mind – the self or soul – is a spin doctor, not the commander-in-chief.'[30] The mind is constructing a story that fits with decisions after they have been made.

As I write these heady sentences, I pause and pick up my coffee mug. This simple act is one of nature's miracles. First, who made that decision if not me? More disturbingly, how can my mental thought cause my physical hand to move? How does mind interact with body? These are some of the most profound issues that have preoccupied thinkers for millennia, but most of us never even bother to consider how amazing these questions are. This is because we do not see a problem at all. We treat the mind and the body as separate because that is what we experience. I am controlling my body, but I am more than just my body. We sense that we exist independently of our bodies.

For most of us, if feels as if we spend our mental life somewhere resident behind our eyes, inside our heads. If we want to see what is behind us, we steer the ship around in order to look. If we want the coffee, we engage the coffee acquisition mechanisms. We feel like pilots controlling a complicated meat machine. There is only one Numskull in control inside my head, and it is I. But how can a nonphysical me control the physical body? How can a ghost inside my head pull the levers?

The dualist philosopher René Descartes proposed that the mental world must control the physical one through the pineal gland deep in the middle of the brain, which he called the seat of the soul.[31] Descartes's solution represents dualism, which requires that there be a soul that is separate from the body and yet in control of the body. But substance dualism must be wrong. The mind is not separate from the body but rather a product of that three-pound lump of gray porridge in our heads. When you damage, remove, stimulate, probe, deactivate, drug, or simply bash the brain, the mind is altered accordingly. In the last century, the great Canadian brain surgeon Wilder Penfield pioneered operations on awake patients for the treatment of epilepsy, including his own sister.

He would expose the surface of the brain and then stimulate the region he was about to operate on to make sure that he was avoiding motor areas that might leave the patient paralysed. When he stimulated the brain directly, the patients experienced movements, sensations, and vivid memories. They tasted tastes, smelled smells, and relived past experiences. Direct stimulation proved that mental life is a product of the physical brain.

Even if there were a seat of the soul that controls our body, how could we explain the relation between these two types of substance, the one immaterial and the other material? In other words, how could an immaterial substance act on a material one? It's not clear how this could work. Descartes's way of solving the mind–body problem, by suggesting a soul that controls the body through the pineal gland, crosses the boundaries between what we know about mental states (that they are immaterial) and what we know about physical states (that they are material). If something nonmaterial could act directly on something material, this would require a mechanism beyond our natural understanding. It would have to be supernatural.

And yet this is exactly what all of us experience on a daily basis. We don't just believe that we are different from our bodies, but rather, as Bloom points out, that we occupy them, we possess them, we own them. Again, this is an illusion that the brain creates for us. For example, when you cut yourself, you feel the pain in your finger, but it is in fact in your brain. When you take a painkiller, it works by altering the chemistry of your brain, not your finger. And yet you feel pain in your finger. Patients unfortunate enough to lose a leg or an arm through amputation often experience their missing limb.[32] Just like real limbs, these 'phantom limbs' get itchy and can be tickled, but they too are an illusion. They are a product of a brain that has failed to update the loss of a body part in its overall map. As if some controller Numskull is looking at the schematic for the factory floor and has not realized that one section has been closed down. The brain areas previously responsible for receiving signals from the missing limb continue to fire away as if the limb were still connected. These examples prove something very disturbing. The brain creates both the mind and the body we experience. A physical thing creates the mental world we inhabit.

This experience of mind is personal and unavoidable. The Harvard psychologist Dan Wegner thinks that the experience of conscious free will in our minds may work like Damasio's emotional somatic marker.[33] Remember how emotions help us in our decision-making by giving us a sense of certainty? Wegner thinks that the experience of conscious free will works in a similar way. My body may tell me that it wants a slurp of coffee, but I experience the decision as my desire to have a drink. This enables me to keep track of my decisions by enriching them with a feeling of control. This is why we have the experience of purposeful decision-making and conscious appraisal. We need to take note of events for future reference. But we would be wrong in assuming that our mental experience at the time is responsible for the decisions we make.

Is all human mental life like this? What about plans for the future, such as schemes for revenge, humanitarian goals, and the need to crack jokes or write popular science books? In what sense could a conscious automaton be responsible for the whole gamut of mental life and aspirations that seems aimed at a future that has not yet happened? The fact that human activities and mental experiences are complicated is not under question. But in the same way we look at complex structures or behaviours in the animal world, such as building a spiderweb or constructing a wasp nest, and wonder how things so complicated could have evolved in creatures to which we don't attribute minds, then we must equally entertain the possibility that humans are just more sophisticated life forms – forms that are capable of making plans and anticipating outcomes. The factors that feed into these processes that lead to complex mental lives in humans are diverse and multifaceted. The mental experiences that accompany such processing are undeniable, but we don't need to evoke a mind that exists independent of and separate from the physical brain to explain them.

Even as a scientist aware of the problem of substance dualism and why Descartes's solution is necessarily wrong, I still cannot ignore the overwhelming sense of my own mind as separate from my body and in control of my body, but ultimately I know it is a product of my body. How do the two interact? That's the mind–body problem. That's what keeps me awake at night. If all my daily conscious experience of a 'me' residing in my head like a Numskull boss were actually true, then it would require a supernatural explanation to make sense of it. That's because we have no natural explanation of how something that has no physical dimensions can produce changes in the physical world. This is why the mind–body problem is one of life's great mysteries.

MIND MY BRAIN

The mind–body problem simply does not appear on most people's radar. It is not a problem until someone points it out or you read books like this one. People have a vague notion that the mind and brain are somehow linked, but rarely do they stop to ponder how the two could actually talk to each other or how something nonphysical could interact with something physical. Most humans have experienced the consciousness of their own minds from an early age, even before they discovered they had a brain. Therefore, it's not surprising that young children can tell you more about their mind than they can tell you about their brain.[34] However, they rarely use the word 'mind', but rather use 'me', 'my', and 'mine'. It's a natural way to describe oneself. The brain, on the other hand, is something they have to learn about, and that comes with science education.

You can find out how much children learn about the brain by asking them a series of 'Do you need your brain to . . . ?' type questions. By the first year of school, most children, like adults, understand that brains are for thinking, knowing, being smart, and remembering. However, they still feel they have a mind in control of and separate from their brain. For example, they do not regard the brain as being responsible for feelings such as hunger, sleepiness, sadness, and fear. From the child's viewpoint, 'It is me who is sad, me who is tired, and me who is hungry.'

These responses tell us that children regard feelings as more personal than thoughts. This is because feelings affect us in a direct emotional way. When we are sad, we feel the pain, the misery, or the despair. It is 'me' that suffers. When we are happy, we feel the elation, the excitement, or the contentment. Feelings are like an emotional barometer of change that we can compare from one moment to the next. It makes a lot more intuitive sense to say that I am a lot happier than I was yesterday than to say that my body and brain are producing different types of mood experiences from one day to the next.

More telling of children's dualism is the way they consider the origin of actions. Actions are controlled by the mind. So kicking a ball or wiggling my toes is a decision made by me, not by my brain. These sorts of answers reveal that children are indeed intuitive dualists. When asked, 'Can you have a mind without a brain?' all six- to seven-year-olds said yes. Science education does little to alter this belief: most fourteen- to fifteen-year-olds agree that the mind does not depend on the brain.

My hunch is that most adults also think the mind can exist without the brain. They may know the scientific position that the mind is a product of the brain, but as we saw with people's understanding of natural selection, knowing the correct answer does not make it feel right. Adults who accept that the mind depends on the brain are likely to still make the same mistake as Descartes in thinking that the immaterial mind acts directly on the material brain.

ROBOCOP

When Officer Murphy was terminally wounded in the sci-fi film *Robocop*, he underwent radical reconstructive surgery to make him into a powerful cyborg.[35] His brain survived, but his memories were wiped clean so that he could become Robocop. His colleagues treated Robocop as a machine, but his former partner, Officer Lewis, detected that there was still something of Murphy present. Over the course of the movie, the cyborg eventually regains traces of his memory to become Officer Murphy again. This tale of human identity is a familiar theme in fiction.

A travelling salesman wakes up to find himself transformed into a giant verminous bug in Kafka's *The Metamorphosis*, but he is still Gregor Samsa because he has Gregor Samsa's mind. The replicant in the sci-fi modern classic *Blade Runner* is convinced that she is human because she has childhood memories, but the Tyrell Corporation, which created her, also fabricated her childhood.[36] It would appear that the hallmark of human identity is a mind full of memories. Maybe that's why most people say they would save a family album full of recorded memories from a burning house.

These examples suggest that we have some strong opinions about what makes something a unique human person, and they make for some interesting thought experiments.[37] For example, imagine that Jim is involved in a terrible car crash and ends up in the hospital, where all the doctors can do is to offer a brain transplant. Consider these different scenarios. Jim's brain is transplanted into a human donor's body. Jim's brain is transplanted into a donor body, but his memory is accidentally wiped clean during the operation. Or Jim's brain is transplanted into a highly sophisticated cybernetic body. After the transplantation, Jim's original body is destroyed. Which, if any, of these patients is still Jim?

Adults are more likely to say that Jim is still Jim if his memories are left intact, irrespective of whether his brain ends up in a human donor body or an artificial cybernetic one. Our conscious experience of our own minds and memories inclines us to think of minds being unique and the source of personal identity. We certainly don't think our own minds and memories could belong to other people. So Jim is like Officer Murphy. He is the product of his mind and memories, and if these can be transplanted, even into an artificial body, he remains Jim. However, the patient with the brain but no memories is deemed to be more human than the cybernetic body containing Jim's brain with memories. This pattern reveals that people consider humans in terms of a physical body and a unique mind that can exist separately.

What about minds existing independently of brains? Most laypeople think that the mind is separate from the brain. After all, the majority of humans have lived their lives never knowing that they possessed a brain, let alone knowing what it might be useful for. Also, as we will see later, people think that it might be possible to copy a body through some form of technology, and possibly even duplicate a brain, but they are less likely to think that a mind could be similarly copied. Moreover, if we could download the mind into another brain, most people assume that the identity associated with that brain would also change with the new mind. So we are naturally inclined to see minds as unique identities that can exist independently of the brain. If this distinction is drawn from an early age, it is easy to see how it leads us to the position that minds are not necessarily tethered to the physical brain. If this is so, then the mind is not subject to the same destiny as our physical bodies. Such reasoning allows us to entertain the possibility that the mind can outlive the body.

AFTERLIFE

In my experience, most Western parents don't talk with their children about death unless they are comfortable with religious explanations. As someone who does not believe in an afterlife, I have found it very difficult to discuss death with my young daughters. It's too painful and awkward. To begin with, you don't have a happy ending, as you do with religion. Also, by discussing death, you are acknowledging that we are all destined to die one day. I will die, and my children will die. It's the ultimate separation anxiety for both parent and child. This makes for a very uncomfortable reality check. All those oxytocin moments seem hollow, artificial, and ultimately pointless when faced with the prospect of death. I would imagine that most atheist parents like me probably avoid discussing death with their children to spare them the difficulty in coming to terms with an existence that has no purpose.

So young children are understandably confused by death. They do not know that all life comes to an end. They do not know that they are going to die one day. They do not appreciate that death is inevitable, universal, irreversible, and final.[38] There are two main reasons for this. First, they cannot conceive of death because they lack a mature

145

understanding of the biological cycle of life and death. As we saw earlier in discussing creationism, children conceive of life as always existing. Second, because of their intuitive dualism, they conceive of death in psychological terms and, in doing so, they can't imagine themselves being dead. So death is understood as the continued existence of the individual, but somewhere else.

Most pre-schoolers think that death is like buying a one-way ticket to a new address with no prospect of return or home visits. When Grandpa has moved on, he has gone to another place. Even if the address is heaven, at least he still exists somewhere. Or they think that death is like sleeping. Certainly ideas of 'departing', 'passing over', and 'resting in peace' are culturally acceptable to tell children and conceptually easier to grasp. No wonder the practice of burying someone in a box under the ground is a very disturbing notion for many pre-schoolers.

When pre-school children were asked in a 2004 study about a mouse that had been killed and eaten by an alligator, they agreed that the brain was dead, but they thought the mind was still active.[39] They understood that bodily functions like the need to eat and drink would stop, but most thought the mouse would still be frightened, feel hungry, and want to go home. Even adults who classified themselves as extinctivists – those who think the soul dies when the body does – said that a person killed in a car crash would know that he was dead.[40] Our rampant dualism betrays our ability to understand that body and mind are tethered together in an inseparable union. When our body packs up, so should our mind. We cannot know we are dead.

Only as children start to learn about what makes something alive do they begin to understand the opposite process of what makes something dead. As we will see in the next chapter, a grounding in biology emerges late in development, and only then do children start to appreciate the mechanics of death.[41] But understanding the mechanics and inevitability of death does not get rid of the belief in the immortal soul. Religion and secular supernaturalism encourage such beliefs, but we must recognize that the concept of the immortal soul originates in the normal reasoning processes of every child. For example, children raised in a secular environment may express fewer afterlife beliefs than children raised in a religious household, but they still retain notions of some form of mental life that survives death.[42] We do not need to indoctrinate our children with such ideas for them to persist.[43] It appeals to our supersense to think that we can continue to exist after our deaths.

WHAT NEXT?

Neuroscience tells us that the physical brain creates the mind. Our rich mental experiences, the sensations, perceptions, emotions, and thoughts that motivate us to do anything, are patterns and exchanges of chemical signals in the complex information-processing of a biological machine. But the mind has no real existence substantiated in the physical world. Psychology is the scientific study of the mind, but the mind does not exist in any material sense. Rather, the mind is the natural operating system that runs on the input and output of the brain's activity. We can study its operations, but we would be wrong to think that the mind occupies a material existence independently of the brain.

However, that's not what we experience when we consider ourselves. We are real, and we exist in the real world. When we think of 'I', we do so in terms of our mind. We experience our mind as an individual motivated by beliefs, desires, emotions, regrets about the past, concerns about the present, and plans for the future. We experience our mind as occupying the machine we call our body. We see our bodies as structures that can deteriorate but we rarely see the structure of our own minds. Even after mental illness, periods of delusion, or temporary intoxication, we usually explain changes in our mind as a result of 'not being ourselves'. This is because we are our minds. The body does not create us. Rather, we are the one who controls it. The philosophical position of substance dualism is the natural way to experience our conscious mind as distinct and separate from our bodies.

Some consider mind–body dualism irrefutable evidence for why there must be supernatural powers operating in the world. The mind is seen as the causal agent but, for that to be true, the mental must be capable of controlling the physical. That would require supernatural powers, since such an arrangment would violate the ontological boundary between the mental and the physical. How else could nonmaterial minds control material bodies? However, most of us don't recognize this position as dependent on the supernatural because minds controlling bodies is the intuitive default of our developing mind-reading of others, as well as our natural experience of our own minds.

The scientific position on substance dualism is that there is no separation of mind and body. It's an illusion as false as the invisible square we saw earlier. Humans are conscious automata. Our bodies generate our minds. When our body dies, so does our mind. But the conscious automation theory is both too unnatural and too repulsive to be accepted by most people. Furthermore, the impression that we have voluntary free will operating within our minds may also be an illusion. Free will requires someone or some ghost inside our heads making the decisions, and that simply gets us into an endless loop. Who is inside our head, and so on, and so on?

So the natural position, based on personal experience, is to assume a separate mind inside the body and not to worry about how the immaterial could control the material. Once we buy into the independent existence of mind and body, there is no limit to what the mind can do. If the mind is separate to the body, it is not constrained by the same laws that govern the physical world. It can leap great distances, travel through solid walls, never age, and travel forwards and backwards in time. In short, misconceiving the mind lays the foundation for many of the beliefs in both religious and secular supernaturalism. In the next chapter, we examine how misconceiving bodies also prepares the ground for our supersense.

CHAPTER SIX

FREAK ACCIDENTS

ON 4 DECEMBER 1980, Stella Walsh, an innocent bystander, was accidentally caught in the crossfire of an attempted armed robbery at a discount store in Cleveland, Ohio. In her day, Stella had been the top athlete in women's field and track events, setting twenty world records and winning both silver and gold medals for the 100-metre sprint in the 1932 and 1936 Olympics. Although resident in the United States, she represented her native Poland in the Games and was awarded that country's highest civilian medal, the Cross of Merit. Everywhere she went, huge crowds turned out to celebrate her victories. In 1975 Stella was inducted into the US Track and Field Hall of Fame. Five years later, a stray bullet in a parking lot ended the life of this once-famous sporting legend.

It was not Stella's tragic death that was to cause a sensation, but rather the results of her autopsy. The sixty-nine-year-old former women's athlete was not exactly who everyone thought she was. She was a he. Despite having been married and living life as a woman, Stella had male genitalia.

Initial reactions to the reports of this discovery led to outraged claims of sporting fraud and cheating. Stella was not a cheat because, technically, she was not entirely a man. She possessed both male and female chromosomes. Stella had a condition known as 'mosaicism', which makes an individual genetically both male and female. Her case was regarded as one of the reasons the International Olympic Committee decided to abandon gender determination tests prior to the 2000 Sydney Games. It's just too difficult to distinguish between males and females, and genitals don't maketh the man.

Mosaics such as Stella Walsh are rare, but it is not their rarity that fascinates us. It was not her sporting fame and untimely death that dominated the headlines at the time, but her being a 'freak'. There are many rare and bizarre medical conditions, but only those that challenge our beliefs about what it means to be a human are called freaks. It's a cruel term that we use to isolate those who do not fit our concepts of what it is to be a human.

During the Victorian era and early 1900s, freak shows were common. In what would now be regarded as politically incorrect entertainment, it was perfectly respectable to pay to see medical oddities. Conjoined twins, bearded ladies, microencephalics, dwarfs, giants, and albinos were all paraded as wonders of nature. Before the advent of modern medicine, many suffered gross disfigurement and physical abnormality through a variety of congenital disorders and progressive diseases, some of which are largely treatable today.

Famous freaks became celebrities, such as Joseph Merrick, 'the Elephant Man', who was a regular on the Victorian London social scene.[1] Others, such as 'Aloa, the Alligator Boy', enjoyed minor fame trawling through the Dust Bowl towns of the American Midwest during the Great Depression.[2] Many of the acts were billed as part-human, part-animal monstrosities. They were abominations who crossed the boundaries between beast and man.

Although freak shows are now long gone, their publicity memorabilia and postcards are still highly collectible today. I keep a small collection as a poignant reminder of how the sensibilities of society have changed. While it may be no longer acceptable to gawk at physical abnormality, modern confessional TV reveals that we are still fascinated by the more deviant members of our society.

Human freaks challenge our view of the living world. We expect people to look a certain way and be a certain size and shape, and individuals who do not fit these expectations are deemed unnatural.

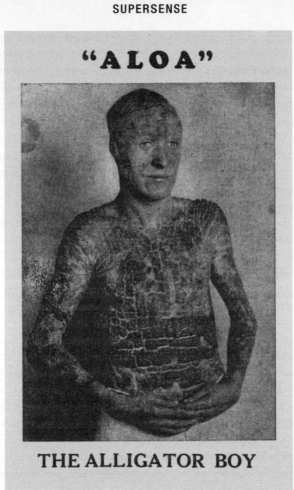

FIG. 12: Aloa, the unfortunate 'Alligator Boy'. AUTHOR'S COLLECTION.

When they have properties that violate our boundaries for grouping the world, they become freaks. For example, bearded ladies, hermaphrodites, and various transsexual combinations contradict our naive biological concepts about what it is to be a man or a woman. Our obsession with genitalia may be motivated by sexual interest, but they are also conspicuous

151

markers for males and females. Whenever genitals are missing, diminished, shared, or on the wrong body, the individual's identity is questionable. Likewise, those endowed with above-average sexual characteristics are judged to be more of a man or more of a woman. Size does matter in this judgement, rather than number. Those unfortunate enough to have multiple penises or vaginas and anything other than two nipples or breasts are generally regarded as freaks.[3]

Where do we get our biological concepts? In this chapter, we are going to look at how the child constructs an understanding of the living world by applying the same intuitive theory-building we saw with minds and objects.[4] Children begin by organizing the world and sorting it into categories. In trying to explain what they observe, they naturally assume that the living world is permeated by invisible life forces, energies, and patterns that define which categories things belong to. This is the stuff that animates matter and makes living things unique. Just like the intuitive theories of the mind that we saw in the last chapter, intuitive biological theories of life lead us to assume a number of ideas that lay the foundation for supernatural thinking.

Like the ancient Greek philosophers, children infer that living things have something special inside that makes them uniquely alive. They assume that there are essences[5] that define what a living thing is, that there are vital life energies[6] that cause things to be alive, and that everything is connected by forces. In philosophy these different but related notions are called 'essentialism', 'vitalism', and 'holism'. As far as they go, they are pretty good approximations of what we know from science about life. If you open any modern biology textbook, you will find that such beliefs are in fact scientifically valid. For example, DNA is a biological mechanism for identity and uniqueness, which are core components of essentialism. Within all living cells there is a chemical reaction known as Krebs's cycle that produces measurable quantities of energy.[7] This is the vital life force that keeps the cell alive. Symbiosis is the study of the interconnectedness of biological systems. The connectedness of living systems can be found in evolutionary theory, in symbiotic physiology, and, more recently, in James Lovelock's 'Gaia' theory of ecology.[8] No man – and for that matter no microbe – is an island; all must be understood as part of a complex system. Most of us are ignorant of these various discoveries and theories, but long before DNA, Krebs's cycle, and symbiosis became mainstream science, humans naturally assumed their existence in the form of intuitive essentialism, vitalism, and holism. However, such intuitive reasoning also forms the core of the supersense because we infer essential, vital, and connected properties operating in the world that go beyond what has been scientifically proven.

Although we intuitively think of essences, life forces, and holism, we would be hard-pressed to describe what we mean. We can't easily articulate these concepts because we often lack the appropriate terms or language. In Eastern cultures, such notions are recognized by ancient terms such as '*chi*' (Chinese), '*ki*' (Japanese), and '*mana*' (Polynesian). In Europe we used to have the term '*élan vital*' (life force), but this has been mostly abandoned. Having good or bad 'vibes' is the closest that most of us come to phrasing these concepts. We may have lost our words to describe them, but our behaviour and opinions reveal that essentialism, vitalism, and holism are still guiding our reasoning. When people respond negatively to wearing a killer's cardigan, this is a reflection of their naive biological reasoning at work. The evil they think is imbued in the cloth is a reflection of the same mechanisms that children apply to infer the hidden properties of living things.

If such metaphysical beliefs are rarely discussed in the West and no one told us about them, then where did they come from? Once again, the most likely explanation can be found in the developing mind. They must come from our natural way of reasoning about life. In this way, children's intuitive biology sows the seeds of adults' supernaturalism, especially as our understanding about life influences much of our attitudes and beliefs.

KOSHER CATEGORIES

Jewish dietary law forbids the consumption of certain animals described in Leviticus of the Old Testament as unclean. At first, the lists seem rather arbitrary. Unclean animals include camels, ostriches, sharks, eels, chameleons, moles, and crocodiles. I have actually eaten three off this list without any ill effects. Some of the animals deemed fit for eating are even more unappetizing to modern tastes, such as gazelles, frogs, grasshoppers, and some locusts. On what basis did someone decide that sharks are unclean but most fish are acceptable? Sharks are fish after all.

Some people have suggested that avoiding certain taboo foods reduces the risk of infection. For example, there is a high risk of food poisoning from shellfish, which can spoil rapidly in hot climates. Undercooked pork can be a source of the parasitic infection trichinosis. However, such an explanation fails to account for many of the unclean animals.

One intriguing alternative is that originally the animals were deemed either clean or unclean depending on how well they fit properties of the group to which they belonged.[9] In the case of mammals, it is clear that the clean or unclean judgement had something to do with how well each example fit general categories when it came to hooves and chewing the cud.

> But this is what you shall not eat from among those that bring up their cud or that have split hooves; the camel, for it brings up its cud, but its hoof is not split – it is unclean to you; and the hyrax, for it brings up its cud, but its hoof is not split – it is unclean to you; and the hare, for it brings up its cud, but its hoof is not split – it is unclean to you; and the pig, for its hoof is split and its hoof is completely separated, yet it does not chew its cud – it is unclean to you. You shall not eat of their flesh nor shall you touch their carcass – they are unclean to you.
>
> Leviticus 11:4–8

Any group of animals should share more properties compared to those from another. Biologists call this grouping 'taxonomy', after the Greek *taxis*, which referred to the main divisions of the ancient army. Today's modern taxonomy is based on one originally devised by the Swedish biologist Carl Linnaeus in the eighteenth century, but prior to this, taxonomies were based on animals' different modes of movement and their habitats.

All the various animals of the land, sea, and air share very similar bodily structures and forms of locomotion. Land animals have four legs and jump or walk. Fish have scales and swim. Birds have wings and fly. One suggestion is that unclean animals tend to be those that violate these properties of the general category to which they belong. Sharks and eels live in the sea, but do not have scales. Ostriches are birds, but do not fly. Crocodiles have legs that look like hands. Maybe some of the unclean animals are the freaks of their taxonomic group. The early Jewish scholars thought that such violations were abominations of the natural world.

Our inclination to understand the world leads us to chop it up into all the different categories we think exist. By looking for the structure in the natural world, we group natural things together into their various kinds. In doing so, we acknowledge that members of a group share the majority of characteristics compared to members of a different group. However, in categorizing the natural world, we become aware that some members do not fit neatly into one category or another. Unclean animals and human freaks are violations of the natural order of things, and that order is one that we construct as part of the intuitive biology we develop as children.

IS IT A BIRD? IS IT A PLANE?

Give a twelve-month-old infant a bunch of toy birds and planes to play with. Then sit back and watch as something quite extraordinary happens. After the initial examination with eyes and then mouth, the baby will start to touch each of the birds in sequence, followed by touching each of the planes. Even though they may have similar shapes, with long bodies and stuck-out wings, the infant is treating birds and planes as different types of things.[10] More remarkable is that six-month-olds shown different pictures of cats and dogs can tell the difference even though no two animals look alike.[11] This simple demonstration reveals some very important things about babies. For a start, they are naturally inclined to sort out the world. They are thinking about things and forming categories. They must be thinking, *This is one type of thing, whereas that is another.* It's exactly the sort of observational technique that professional scientists use when trying to understand the world. By sorting, they are telling us that they understand that dogs are members of one category whereas cats belong to another. In short, they have a rudimentary biology.

When and where does the child's understanding of biology come from? The Harvard psychologist Susan Carey argues that children take a relatively long time to understand biology. They may be able to sort birds and planes and cats and dogs, but Carey thinks that such categorizing is only simple pattern detection that doesn't require a deep understanding of biology. To get to grips with biology you have to appreciate life as a state of being, as well as the invisible processes associated with it. In Carey's reckoning, it's not until age six or seven that children begin to understand what it is to be alive.[12]

Also, babies may spot the difference between living and non-living things, but they could just be making judgements based on how humanlike something is. In other words, they may be thinking nothing more than that the closer something in the natural world is to looking or behaving like a human, the more likely it is to have the same biological properties as humans. It's anthropomorphism at work again, not reasoning about other life forms as separate categories. We can get an idea of children's level of biological knowledge if we show them pictures of plants, insects, animals, and objects and ask them questions such as: Does it eat? Does it breathe? Does it sleep? Does it have babies? The closer things are to looking or behaving like humans, the more biological properties children give them. For example, in one study pre-schoolers thought that dogs and even mechanical monkeys were more likely to eat, breathe, sleep, and have babies in comparison to bees and buttercups because they resembled humans and seemed more purposeful than insects and plants.[13]

As far as it goes, this is not a bad strategy. It's the pre-schoolers' equivalent to the 'if it looks like a duck, waddles like a duck, and quacks like a duck, then it's probably a duck' approach to figuring out the living world. However, more recent research suggests that pre-schoolers do have something akin to a biological awareness that goes beyond simple outward appearances. Children think that there must be something inside animals that makes them both unique and alive. Before they reach school, children start to think like most adults in terms of essences and life forces. They are intuitive essentialists and vitalists.

THE ESSENCE OF LIFE

What is an essence? Consider the real essence of physical chemical compounds. Both flowers and cats can produce such a physical essence. In perfumery, essences are the concentrated reduced quantity of a fragrant substance after all the impurities have been removed. Chanel No. 5 is one of the world's most successful perfumes and is very expensive because of the cost of harvesting one of its chief ingredients, jasmine blossoms. These are grown in the Provence region of France and survive for only the briefest time before losing their fragrance.

Another reason Chanel No. 5 is expensive, apart from jasmine essence, is that until recently it also contained musk secretions from the anal glands of the endangered civet cat, the same species in Asia that excretes coffee beans to produce the Kopi Luwak gourmet coffee mentioned earlier. (The civet cat is not actually a cat but a raccoonlike creature.) Musk is a sex chemical that a number of mammals use to attract partners and mark their territory. The pungent smell takes a long time to decay, and so perfumers use musk to prolong the scent of more fragile fragrances. When it became widely known that Chanel used civet cat musk in its perfumes, Chanel replaced this ingredient with a synthetic musk compound. It is not clear whether this decision was due to pressure from the animal rights groups concerned about the cruelty inflicted during the musk extraction process or, more likely, to consumers' distaste at discovering that they had been smearing secretions from an animal's bum around their delicate wrists and necks.

In philosophy, essences are less smelly. In fact, you can't detect them at all because they exist beyond man's ability to perceive. Greek philosophers thought essences were some inner, invisible substance that made things what they truly were, like another dimension to reality. For example, Plato, probably the most prominent exponent of essentialism in his theory of ideal forms, argued that everything has an inner reality that we cannot necessarily perceive. Aware that appearances can be deceptive, he proposed that the world we experience is only a shadow of true reality. He likened human experience to sitting in a cave and watching reflections of reality from outside projected as shadows on the cave wall. It's a bit like our *Matrix* comparison again. We glimpse only a fraction of the reality that truly exists. Plato thought that humans could never get at the true essence or form of things because of the limits of our minds.

Plato's analogy is true in some sense – well, actually, all senses when it comes down to it. Our brains can process only the information we receive from the outside world through our senses. But our senses are limited. We know there is sound we cannot hear, light we cannot see, smell we cannot detect, and so on.[14] This means that there are things in the world that we cannot directly perceive. There are microbes, viruses, particles, atoms, and all manner of teeny-weeny things that we know must exist but that are invisible to us. We are only ever glimpsing a portion of reality. Likewise, early essentialists thought that essences reside beyond our sensory range. Plato thought each essence is the core internal property that gives a thing its unique identity.

An essence is not to be confused with any unique property. For example, humans are the only mammals that have opposable thumbs. Thumbs may be unique to humans, but they are not essential. You would still be human if you were born without thumbs. Rather human essence is some invisible property that distinguishes us from non-humans. Like the pod-people in the sci-fi classic *Invasion of the Body Snatchers*, alien replicants might be identical to us in every physical way, but they would lack the essential quality that makes us human.[15]

As comforting a notion as human essence might be – that even though our bodies wither and decay there is some enduring stuff inside us – this philosophical position is a logical nonstarter. That's because there is more than one way to define any object, including a human. The same individual human can simultaneously be a male, an adolescent, a prince, a neurotic, an artist, an athlete, an atheist, and so on. An object can be a stone, a paperweight, an ashtray, a weapon, a counterweight, or even a sculpture. And if there is more than one way to define an individual, you can't have a unique essence of that individual. Aristotle was Plato's student, but he realized that his teacher had been mistaken as far as essences were concerned. So the idea that there is only one true individual essence is nonsense.

When art critics and gallery owners talk about the essence of a piece of art, they are talking essential nonsense. However, just because something is nonsense doesn't stop people believing in it. People can still hold a psychological essentialism.[16] It helps us to think about uniqueness as a tangible property. This is my cup. This is my Picasso. This is my body. Psychological essentialism is the *belief* that some individual things, such as other people or works of art, are defined by a unique essence; as we will see in the coming chapters, such a belief would explain many of our attitudes when we think essences have been violated, manipulated, duplicated, exchanged, or generally mucked about with. Humans like to think that special things are unique by virtue of something deep and irreplaceable. When we chop nature up into all its different groups of living things, we are assuming that these are groups of things that are essentially different.

THE ESSENTIAL CHILD

Children's essentialist thinking is amazing.[17] Before they reach school age, they know that baby joeys raised by goats grow up into adult kangaroos, not adult goats. They know that apple seeds grown in flowerpots become apple trees, not flowers.[18] They even know that a light-skinned baby switched at birth with a dark-skinned baby remains the original colour despite being raised by the new family.[19] A leaf insect may look more like a leaf than an insect, but four-year-olds know it shares properties with other bugs, not with leaves.[20] When they are slightly older, they understand that if an evil scientist takes a raccoon and performs an operation to turn it into a skunk by attaching a furry tail, painting a white line down its back, and putting a bag of foul-smelling stuff between its legs, it is still a raccoon even though it looks like Pepé Le Pew.[21] Essential thinking allows children to understand that the leopard literally can't change its spots. And no one needs to teach children this. It's part of their intuitive biological understanding.

Children's essentialism is truly surprising, as pre-schoolers can often be fooled by outward appearances.[22] However, once they understand what can and can't be changed by environment, they are committed essentialists who see core properties everywhere. They think that there is something inside that cannot be changed. They don't know what it is, and they would be hard-pressed to describe it. When it comes to understanding living things, they really seem to grasp that there is something deep inside that makes animals and plants what they are. It's a universal beliefs shared by different cultures, suggesting that essentialism is a natural way of viewing the world.

Although children and most adults can't describe exactly what an essence is, they can tell you where it is, if only indirectly. In one study, children were told about an ancient block of ice that had different animals frozen in it.[23] Scientists wanted to determine what the different animals were by doing tests on small samples taken from the things inside the block. Children were asked whether it made a difference where the sample was taken from. By ten years of age, children reasoned like adults that it did not matter where the sample was taken because whatever defines an animal is spread throughout the body. In contrast, four-year-olds, the youngest children in the study, insisted that the true identity of an animal is found in only one spot and not spread out. When questioned further, these children seemed to think that the correct spot to choose was from the centre of the body. What starts out as a very localized notion of essence in young children develops into a belief about something that spreads throughout the body, even though these children never mentioned scientific concepts such as DNA.

POLAR MICE AND FISHY POTATOES

Essential thinking is increasingly shaping our attitudes towards the modern world. For example, by the time the leaves on a potato plant start to wilt, the potatoes underground are already stunted in size as the plant tries to compensate for lack of water. What if the plant could tell you that it needs watering before the leaves begin to wilt? There is one such potato plant whose leaves start to glow fluorescent green when they require water. It can warn you in advance that it needs water before the underground potatoes shrivel. The plant can do this because a gene from a jellyfish has been inserted into its genetic makeup. It's a genetically modified plant. When water levels reach the critical level, the gene in the plant's physiology turns on the fluorescent response. A potato that can communicate its needs is truly remarkable – almost sociable. But would you eat such a fishy potato?[24]

Or what about a supermouse that can survive freezing temperatures? The Alaskan flounder produces a protein that effectively produces an antifreeze in its blood to enable it to survive in subfreezing waters. Last year mice were bred with this gene that protected them from hypothermia.[25] What's more, the mice passed this gene on to their babies, demonstrating the potential to create new species of animals that cross the traditional taxonomic boundaries. In other words, these supermice were genetic freaks.

Those biological boundaries that we use to chop up the world are increasingly open to breech by new genetic engineering. There are real concerns about this technology, as it is not easy to predict exactly what unforeseen negative consequences may arise from artificially combining genetic material that would not normally occur in nature. In the remake of the sci-fi classic *The Fly*, the scientist Seth Brundle builds a machine that decomposes the body down into its constituent DNA particles and transports them from one pod to another where they are reassembled.[26] By chance during one of his early experiments, a common housefly enters the pod with Seth. At first he notices nothing when he re-emerges from the other pod, but over the course of the movie

Seth is gradually transformed into a human fly hybrid, with all the disgusting dining habits that flies exhibit (and you know what I think about flies). In most people's minds, genetic engineering has brought us to the point where Seth Brundle's predicament is no longer a fanciful tale of the dangers of tinkering with nature.

It's not the fact that we can do genetic manipulation that is so worrying. After all, from the very beginnings of farming and animal rearing, we have been manipulating genes through selected breeding. All modern dogs are descendants of a fifteen-thousand-year-old programme of selective breeding of wolves.[27] The problem is that gene insertion rapidly bypasses natural selection. There is no time to evaluate combinations that could be harmful. The potential for unforeseen consequences arising from unconstrained combinations worries the experts.

Around the world, governments are anxiously weighing up the concerns raised by genetic engineering with the potential benefits of new solutions to problems. For example, stem cells are the juvenile cells in fetuses that have the potential to replace damaged cells in adults.[28] Many people suffering from illnesses and diseases such as Alzheimer's disease would benefit if stem cells could achieve this repair. Unfortunately, there are not enough spare human eggs available to conduct this research, and so one solution has been to create animal eggs containing almost entirely human DNA. The resultant embryo, however, would still contain a small proportion of the donor animal's original genetic material. This hybrid human–animal embryo could in principle be a potential real Seth Brundle. In truth, these embryos would never be viable, but the prospect of animal–human hybrids is simply too unacceptable for most of us. In March 2008, the British government faced a crisis as the Catholic Church urged Catholic politicians to resign over the introduction of the Human Fertilization and Embryology Bill, which allowed research inserting human DNA into animal cells. Ethics used to be a rather sleepy academic division of moral philosophy where one could ponder life's hypotheticals. Today, advances in genetic engineering have thrust ethics into the public spotlight, with the expectation that it will provide answers to this minefield of moral dilemma. Philosophers have never been busier.

Your average member of the public has never taken courses in philosophy or genetics, but they can still be appalled by the prospect of combining species. This is because of essentialism. It's the way we all chop up the living world into its different groups. We intuitively think that members of the same category share this invisible property that defines their group membership. For example, we think that all dogs have a 'dogginess' essence that makes them members of the canine family and that all cats have a 'cattiness' essence that separates them from dogs and makes them members of the feline fraternity. When we hear about scientists inserting genes of fish into mice and potatoes, we feel squeamish. It just does not seem right. It's not natural.

Who did not feel the 'yuck' factor when they first saw the picture of a hairless mouse with what looked like a human ear growing on its

FIG. 13: Mouse with an implanted bioframe. Many people have mis-interpreted this image as an example of genetic engineering.
© BRITISH BROADCASTING CORPORATION.

back, circulated around the world's media? It wasn't actually an example of gene manipulation but rather a demonstration of how an animal could be a surrogate host for growing an implanted bioframe.[29] But it certainly looked like a hu-mouse! Our revulsion was not simply because it was a weird image. Rather, we felt simultaneously sick and fascinated because the prospect of human–animal hybrids violates the essentialist view of the world that we developed naturally as children. When I was preparing this chapter, my youngest daughter looked over my shoulder and saw the image of the mouse with the human ear. At first she let out an audible 'yugh!' Then she asked if it could hear better. Apparently she is still telling her classmates about it.

MAY THE FORCE BE WITH YOU

Related to the notion of essence is the idea of a life force, something that is in living animals but not in dead ones. This is vitalism, an ancient belief that the body is motivated by an inner energy. Up until the nineteenth century, this was recognized in the West as the '*élan vital*', a vital force that does not obey the known laws of physics and chemistry.[30] In most conceptions of a life force, it is equated with the unique identity of the individual. In other words, it is the essential soul that many believe inhabits our bodies but departs on death to move on to another dimension/body/location/time (delete terms as appropriate depending on your afterlife belief system).

Although we cannot see the energy generated in our bodies, most of us intuitively feel it is there – not within every living cell, as Krebs's cycle describes it, but rather as a unified whole thing that animates the body. I can relate to why people think this. On occasion, I have had to kill animals, either for food or because they have become a nuisance. I live in the country and raise my own chickens for the table. When they are ready, I pull their necks. When you kill a largish animal up close, as opposed to squashing a fly with a rolled-up newspaper, you can experience a sense that something leaves the body. A living entity that a moment ago was animated, flapping around and agitated, is now still. But there seems to be more involved than just an absence of movement.

I have seen a number of corpses in the dissection room, but I have not watched someone die in front of me. However, I have talked to friends and colleagues who have been at the bedside of a dying person, and they often report that something seems to depart. So far no one has told me that they actually saw something leave a body. Rather, they get a feeling that someone or something has left. Maybe this is what our minds create in order to makes sense of the change in the situation: suddenly there is one less person in the room. How can there be one person less in the room unless someone has left?

In popular culture, the moment of death is often depicted as a life force or energy leaving the body like some semitransparent copy of the person. This notion may be purely psychological, but there are many people who think a tangible soul exits the body at death.[31] In 1907 Dr Duncan Macdougall of Massachusetts reported that the soul weighs precisely twenty-one grams based on his careful measurement of six dying patients on a set of industrial scales.[32] His findings were and have since been treated with much scepticism, with alternative explanations ranging from fraud to methodological weakness. Because the weight loss was not reliable or replicable, his findings were unscientific. When he was prevented from further human studies, Dr Macdougall moved on to dogs that he sacrificed in his scientific search for the soul. The results of these studies showed no evidence of a weight loss at the time of death. Undeterred, Macdougall interpreted this as evidence for the Christian belief that animals don't have souls. In which case the word 'animal' is inappropriate, as it comes from the Latin *anima*, for soul.

Scientifically, death is another continuous stage of life. At death, the meat machine no longer functions as a unified system and begins to decompose. It starts to disassemble itself. In the absence of oxygen, the cells start to die. Krebs's metabolic cycle shuts down, and the system starts to go into reverse. The bacteria colonies that once helped to sustain life now begin to break the body down. Like opportunistic looters, they requisition various material substances to embark on their own life cycles in isolate. It's like the breakup of an army. Once the battle is over, the individual soldiers take what they can and then head off. The state of death is simply the process of life in different directions. With the defence systems down, all manner of microbe, insect, and beast plunder the body for resources. If we could record and play our lives out as one of those time-lapse movies of decaying fruits and animals, we would realize that composition and decomposition are continuous.

Such an account is neither comforting nor acceptable for most. Where has the person gone in this version? The body remains, but the person is absent. A departing life force that energized the body is the only sensible explanation for most people. The mind–body dualism we intuit when we are alive explains to us what happens when we are dead. And, like dualism, the notion of a vital energy inhabiting the body is a concept that emerges early.

Young children understand life in terms of a vital energy necessary for keeping the body going.[33] In one investigation, children were asked different biological questions, such as, 'Why do we breathe?' To help them answer, the researchers offered the children three types of explanation: those based on mental goals (because we want to feel good), mechanical explanations (because the lungs take in oxygen and change it into useless carbon dioxide), or vitalistic explanations (because our chest takes vital power from the air). By six years of age, most children endorsed the vitalistic reasons, whereas older children and adults selected the mechanistic accounts. Education may have taught them about oxygen and carbon monoxide, but the explanation based on vital energy was the default position of younger children. Some children talked about blood carrying energy to the hands in order to make them move. Education provides us with new frameworks of explanation, but as we saw with naive theories of gravity and other intuitive models of the world, it's not clear that earlier ways of thinking are abandoned. An enduring vital force seems a plausible explanation for life.

The concept of enduring life energy is not entirely flaky. A living body does generate energy in that it converts energy from one source into another. This is what metabolism is. Energy is never lost. This is the first law of thermodynamics, discovered over the last three hundred years. Energy cannot be lost but rather changes state. While very few of us are knowledgeable about the laws of thermodynamics, for many the transition from life to death is simply the movement of an energy source from one state to another. Many adults who are ignorant of the biological facts regarding metabolism and energy can nevertheless still conceive of some force that resides in a living thing but moves on at the point of death. We are intuitive vitalists.

But children do not start off as vitalists. The questions confuse them because they have not yet begun to think about their own bodies as separate from their minds. This may explain why they have a problem understanding death, as we saw in the last chapter. When five-year-old children were sorted into those who thought in terms of vital life forces and those who did not, the vitalist children were the ones who understood that death is irreversible, inevitable, and universal and applies only to living things.[34] Younger, novitalist children were just confused. So an emerging naive vitalism helps children to appreciate the nature of death as final and something that happens to everyone. Intuitive theories don't have to be scientifically accurate to be useful.

THE GREAT CHAIN OF BEING

The essential life force is not only an intuitive concept found in every child. It is also a belief that has survived thousands of years in different models of the human body, both religious and medical. The ancient Greeks described the essential life force in their humoral theory of how the body works. They believed that a healthy body depends on maintaining the balance of the four vital juices of blood, phlegm, yellow bile, and black bile. However, as these bodily fluids are ultimately perishable, a fifth element, or 'quintessence', is necessary to animate the body with spirit.[35] Today a similar idea is still the core component of traditional Eastern medicine and philosophy, whose treatments and rituals involve manipulating and channelling energy. The Greeks also recognized a holistic concept of life – the doctrine that unseen energies and forces connect everything in the universe. These connections are permanent, so that action on one thing in the universe has consequences further down the chain. The more closely things are connected, the stronger the consequence of action.

Such an idea underpinned the later dominant Western medieval theory of the universe known as God's 'Great Chain of Being'. This

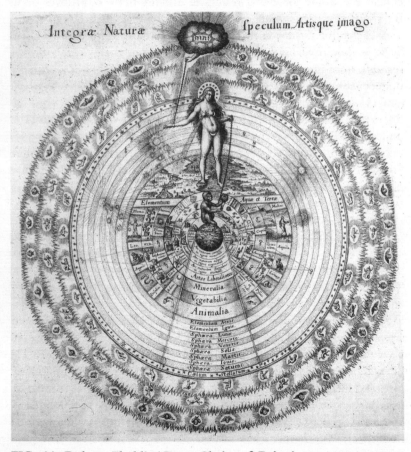

FIG. 14: Robert Fludd's 'Great Chain of Being'. PHOTOGRAPH BILL HEIDRICH © UC BERKELEY, CALIFORNIA

was the belief that all things, including animals, vegetables, and minerals, are related.[36] All things originated from the same source, are organized into a hierarchy of association, and are held together by divine correspondences – invisible forces that connect the various elements. These forces could be sympathetic in that they shared common correspondences that could be combined. Or the forces could be antipathetic where the elements opposed each other and could be used to cancel each other out.

For example, in an illustration of God's natural plan published in 1617, Robert Fludd's diagram shows how man was sympathetically linked to the sun, which was linked to the grape vine, which was linked to the lion, which was linked to gold. Hence, men were noblemen. Gold was considered a noble metal, as was the name of the gold coin of this period. The vine was noble, and the mould that forms on overripe fruit and produces a characteristic rich flavour was known as the 'noble rot'. The lion was a noble beast. Likewise, woman was sympathetically linked to the moon. Her menstrual cycle was clearly related to lunar activity, which was linked to wheat, which was linked to the eagle, which was linked to silver, and so on. Man was opposite to woman, and the sun was opposite to the moon. Everywhere in nature you could find evidence for sympathies and antipathies by looking for signatures of God's hidden order. The evidence was overwhelming. You just had to look around you and see all the connections. This was trivially easy for a human mind designed to detect patterns and infer connections in the natural world.

Everywhere nature's patterns were interpreted as reflecting a deeper causal model based on God's hidden correspondences. Sometimes God left clues in that animals, vegetables, or minerals that shared sympathetic correspondences looked similar. This reasoning became known as the 'Doctrine of Signatures' and was the basis for much alchemy and folk medicine.[37] For example, because walnuts looked like the brain, they were used for headaches. The weeping willow tree was thought to be a cure for melancholy because of the clear signature of the drooping weariness of its branches. The foxglove plant (*digitalis*), with its spotted fingers, was originally thought to be a remedy for respiratory conditions because it was reminiscent of diseased lungs. Turmeric, the root commonly used to colour Indian food yellow, was used to treat jaundice, a condition that produces a yellow skin pallor. Mandrake roots, which resembled shrivelled humans, were considered to be particularly potent and, owing to their alkaloid toxins, could be used to induce altered states of consciousness for all manner of purposes. Nipplewort (*lapsana communis*), a tall weed with small yellow heads, was once esteemed for treating sore nipples. Pilewort (*lesser celandine*), with its knobbly tubers, speaks for itself.

Even today many societies value magical foods that are believed to contain essential healing or enhancing properties by virtue of their resemblance to body parts. Figs and pomegranates have properties that

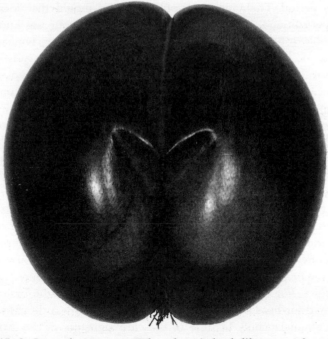

FIG. 15: A Coco de Mer nut. What does it look like to you? AUTHOR'S COLLECTION.

resemble female genitalia. The Coco de Mer coconut resembles a woman's genitals and is highly prized for fertility.[38]

Phallic-shaped foods like bananas and asparagus are also deemed to be potent by virtue of their resemblance to the penis. It's not too surprising then that actual penises feature regularly as foods that can enhance male performance. The Guolizhuang in Beijing is China's first restaurant that specializes in animal penises. Businessmen can pay up to £3,000 to eat tiger penis in the belief that it will improve their virility and life energy.[39]

Much of traditional Chinese medicine is based on essentialist and vitalist notions of sympathies. Pregnant women are advised to eat dragon-tiger-phoenix soup, which combines the energies of snake, chicken, and our old friend the civet cat. Yes, that's right. If it's not enough that we drink its droppings in our coffee and smear its buttock juice on our necks, it's also a popular ingredient in a common Chinese medicinal soup. Civets may have the last laugh against their human tormentors. The severe acute respiratory syndrome (SARS) outbreak that threatened a world-wide pandemic in 2003 was transferred to humans by civet cats stacked in crates in the infamous wet markets of the Far East before being shipped out to restaurants. SARS is a coronavirus. It replicates by hijacking the DNA contents of a cell and replacing it with its own genetic material. You could say that a coronavirus substitutes one essence for another. How ironic that the cherished supernatural essence of infected cats was in fact a real and deadly virile essence with a one-in-ten fatality rate.

HOMEOPATHY IS ESSENTIAL

Modern homeopathy is equally a direct descendant of sympathetic magical reasoning and logic. Much of its practice is based on the publication of the German physician Samuel Hahnemann's (1755–1833) law of similars: *similia, similibus curantur*, or 'like cures like'. If your baby has a diaper rash, homeopathy recommends treating it with poison ivy, a toxin that produces severe rashes. For children's diarrhoea, try a dose of rat poison. But don't worry, the first law of similars was supplemented by the second law of infinitesimals, which states that the more dilute the dose the more effective the treatment.

Homeopathic remedies are diluted to such an extreme that it is unlikely that the liquid contains anything but pure water. This is because the practitioner adds the ingredient to a beaker of water and then takes one-hundredth of the solution and adds this to a new beaker. He or she then takes one-hundredth of that solution and repeats the process over and over again. A typical homeopathic remedy will be so dilute that it contains one particle of the original target ingredient in 1,000,000,000,000,000,000,000,000,000,000,000,000,000,000,000,000, 000,000,000,000 particles of a liquid. You get the picture. You would have to drink twenty-five metric tons of water for there to be a remote chance that you had swallowed just one molecule of the original substance. Apparently this is not a problem. According to homeopathy, shaking the solution ten times with each dilution releases the vital energy of the active ingredient, which imprints a memory trace in the water.

Needless to say, the scientific community regards homeopathy as supernatural quackery.[40] It is based on holist, vitalist, and essentialist beliefs. Yet it is an alternative approach to health that is increasingly popular. In 2007 the United Kingdom's *Times Higher Education Supplement* reported a one-in-three increase in applications to study alternative medicine at alternative educational institutes and a corresponding decline in applications to study anatomy, physiology, and pathology at traditional universities.[41] Homeopathy is available through the National Health Service, and even Bristol is home to one of five NHS homeopathic hospitals, despite the fact that the evidence for the efficacy of homeopathic treatments is at best equivocal. Boots, the United Kingdom's largest chain of pharmacies, once rejected homeopathy but today sells a range of homeopathic remedies. It also includes a full online educational course to teach children about homeopathy, holistic healing, vital forces, and why diluted honey is good for bee stings.

What is it about modern medicine that leads people to prefer to put their faith and the care of their bodies into supernatural remedies?

For one, homeopathy actually works. It works because patients believe it will. On average, one in three sick patients will improve if they believe they are receiving an effective treatment. This is the so-called placebo effect. The placebo effect is the remarkable finding that people get better if they think that they are taking a medicine or undergoing some therapy even if it has no direct active ingredient. Every drug that is regulated in the United Kingdom has to pass clinical trials that prove it is more effective than the results achieved by placebo alone. No such ruling exists for homeopathic treatments. For example, in the United States, Nicorette, a chewing gum that helps smokers give up smoking, had to pass stringent clinical evaluation before its maker gained a license to sell it. But in the same drugstore you can buy CigArrest, the homeopathic equivalent that did not have to pass any such evaluation. It would appear that the regulatory authorities are more concerned about the potential side effects of drugs with active components than treatments that are not distinguishable from pure water. Anyway, how could you prove that any homeopathic remedy did not have the appropriate active ingredient? You couldn't find it if you looked for it!

The placebo effect is very real, and if belief improves health, then should we be concerned by supernaturalism in our health care? After all, homeopathic remedies are just water, and most practitioners refer to them as complementary medicine meant to be used in conjunction with clinically evaluated treatments. If this enhances the placebo effect, so be it. The problem occurs when complementary treatments are believed to be equally effective alternatives. This was revealed in a scandal made public last year about homeopathic anti-malarial treatments. The London School of Tropical Medicine was increasingly alarmed at travellers returning with malaria because they had not taken conventional prophylaxis. They found that of ten randomly selected homeopaths operating in London, all of them recommended taking homeopathic preventive treatments alone.[42] This was despite the recommendation of the United Kingdom's Society of Homeopaths, which concedes that there is no known effective homeopathic anti-malarial treatment.

There must be other reasons why people reject proven modern treatments in preference for supernatural cures. Over the past decades,

173

there has been a change in attitudes toward modern medicine. For one thing, holistic treatments consider the whole of the person, and in doing so alternative therapists spend much more time listening to the patients and their problems in comparison to doctors working to a time-sensitive regime. Patient satisfaction and significant improvement in health are directly related to the amount of time the doctor listens to the patient's problems.[43] Not only is a problem shared a problem halved, but the sharing often leads to significant improvement in health.

Another reason for the rise in the popularity of alternative medicine is that we are increasingly concerned about the advances in science and treatments that seem unnatural. Have you noticed how common the word 'natural' is in advertising today? In our so-called 'postmodern' era, we hanker for a return to a simpler time, and a preference for natural products reflects this changing attitude and anxiety about modern science. But what exactly is a natural cure, and is it less dangerous than modern medical treatments? It turns out that nature has many more natural toxins than those synthesized by man. In fact, much of homeopathy works on the principle of a tiny bit of bad is good for you. So just because a substance is naturally occurring doesn't make it safe.

DISGUSTING RESEARCHERS

The supernatural basis of alternative medicine sounds like the sort of mumbo-jumbo confined to the unenlightened dark ages of prescientific societies. But we should not be so quick to mock those who seek such treatments. The same laws of sympathetic magic are arguably part of daily life for all of us today, and no more so than in the peculiar human experience of disgust and our fears of contamination. Our contamination fears reflect our reluctance to come into physical contact with things that we find disgusting. We may be able to fight the urge and overcome our disgust, but it can operate at a gut level, making it difficult to control through reason.

Some things automatically trigger disgust and don't have to be learned. Hydrogen sulphide, methane, cadaverine, and putrescine are four of the most revolting smells to the human nose. They can be found in various bodily excretions but are most concentrated in a decomposing corpse. When I trod on the stomach of that dead cat as a ten-year-old, it was this chemical cocktail that assaulted my senses. Everyone feels disgusted by the smell of putrefying bodies. However, other triggers of disgust are not so hardwired into our biology, and that is why disgust is so interesting to psychologists: sometimes it can be triggered by belief alone.

When we met Paul Rozin earlier, it was in the context of the killer's cardigan, but this research stems from his work on the origins and development of human disgust. Rozin is one of the most disgusting researchers in the world. After reading about his studies, you would be very wary about stopping over for dinner at his place.[44] For example, he measures how adults respond to various challenges that trigger the 'yuck' response. Could you drink out of a glass after it has been touched with a sterilized cockroach? Could you eat a delicious piece of chocolate fudge shaped like a dog turd? Would you slurp your favourite soup after it had been stirred with a brand-new fly swatter? Why does spitting on your own food make it disgusting despite the fact that you need saliva for digestion? As you would expect, people are disgusted at the prospect of most of these challenges, even though the actual risk of contamination is minimal or nonexistent in each situation.

And then there are cultural variations. Many of us could quite happily tuck into a bacon sandwich (apparently one of the most difficult things for former meat-eaters to give up when they become vegetarian), whereas a devout Arab or Jew would consider it disgusting. In the West, we are appalled at the ease with which insects, penises, gall bladders, snakes, cats, dogs, and monkeys are consumed in the Far East. Clearly some forms of disgust are culturally determined. How can this be?

ESSENTIAL CONTAMINATION

Cultural variations prove that some triggers for disgust must be learned. When we watch others turning up their noses at particular foods or retching at certain sights, we can copy their responses. But disgust and the accompanying fear of contamination do not follow simple learning rules in the normal way. For a start, we are wired to respond automatically to others' disgust. Simply watching someone pull a disgusted expression is sufficient to induce our own feelings of disgust. For example, if you watch somebody pull a face after sniffing a drink, this activates the insula, the same region of your brain that normally fires when you yourself smell something offensive.[45] It's one-trial learning. That's how rapid and infectious disgust emotions can be.

For me, the really interesting aspect of disgust and the associated contamination fears is that they show all the hallmarks of supernatural thinking.[46] This is because they trigger psychological essentialism, vitalistic reasoning, and sympathetic magic. For example, sympathetic magic states that an essence can be transferred on contact and that it continues to exert an influence after that contact has ceased. This is known as the 'once in contact, always in contact' principle.[47] Something you cherish can be ruined by coming into contact with a disgusting contaminant in exactly the same way. For example, the briefest touch of your food by someone you think is disgusting makes the dish unpalatable. There's an old saying that a drop of oil can spoil a barrel of honey, but a drop of honey can't ruin a barrel of oil. This is the negative bias that humans hold when it comes to contamination.[48] We intuitively feel that the integrity of something good can be more easily spoiled by contact with something bad rather than the other way around.

However, it's difficult to be reasonable about contamination once it's occurred. It's as if the contaminant has energy that can spread. For example, imagine that your favourite dessert is cherry pie and that you have the option of choosing between a very large slice and a much smaller piece. Unfortunately, your waiter accidentally touches the crust of the large slice with his dirty thumb. The same thumb that you just saw him pick his nose with. Which slice would you choose? Given the choice, most of us would opt for the smaller slice, even though we could cut off the crust where the waiter touched it and still end up with more pie. As far as we are concerned, the whole slice has been ruined – as well as our appetite.

THE WISDOM OF REPUGNANCE

Disgust affects more than just our attitudes toward the things we put in our mouths. It clouds our moral judgements too. Many people rely on disgust in deciding what they think is right or wrong. Leon Kass, the former chief ethical adviser to President George W. Bush, argued that disgust is a reliable barometer to what we should find morally unacceptable, the so-called knee-jerk response. In his essay 'The Wisdom of Repugnance', he makes the case that disgust reflects deep-seated notions and should be interpreted as evidence for the intrinsically harmful or evil nature of something.[49] If you feel disgusted when you hear about some event, then that's because it is wrong. The problem with this view is that what people find disgusting depends on whom you ask.

Consider incest between a consenting brother and sister. In most societies, brother and sister incest is regarded as disgusting. Why? What is wrong with two genetically related people having sex? We could argue that this response evolved because of the risks of inbreeding. For example, mating with your sibling is a genetic no-no, as there is an increased chance that any offspring would have genetic abnormalities. And yet if a brother and sister have consensual sexual intercourse in private so that no one would ever know, using birth control and basically avoiding any possible chance of pregnancy, we still consider such sex morally unacceptable. Then there are all the other weird things that people might get up to. Why is cleaning the toilet with the national flag or eating a chicken carcass you have just used for masturbation disgusting? These acts might be weird, but there is no intrinsic reason for why they are wrong.[50] What's wrong with wearing a killer's cardigan? In fact, people are often lost for words when trying to give reasons. They are morally dumb-founded, as the psychologist Jonathan Haidt says.[51] By the way, just in case you wondered about the warped state of my own mind, these disturbing examples all come from Haidt's work. So write to him if they upset you.

Biological explanations are too limited for explaining all the things we find disgusting. Rather, the answer must be some other mechanism

177

that uses disgust responses for some other purpose. One possibility is that disgust works as a mechanism for social cohesion. To form a cohesive group we must have sets of rules, beliefs, and practices that define our group and that each member must agree to abide by. This is how one gang distinguishes itself from another. These are the moral codes of conduct found throughout the different cultures of the world. When these rules are violated, a taboo has been broken, and a negative emotional response must be triggered. The perpetrator must feel guilt, and the rest of us must punish that person. This is how justice works. The net effect is to strengthen the cohesion of the group.

Culturally defined taboos may engender social cohesion, but they are not based on any reason other than that they define the group. This is why those individuals who are happy to touch a killer's cardigan are outsiders. By taking a behaviour and linking it to a visceral response, we can use disgust to control individual group members. We can also use disgust to ostracize others. In the next chapter, we examine how such essential thinking is at the root of bigotry directed towards people who some would rather keep at arm's length because of their colour or social background. When we say the peasants are revolting, we mean it in a disgusted way. It provides the emotional reason for treating them the way oppressors do. We can treat others badly who do not share our values because it feels right. And why does it feel right? I think the answer is that a supersense of invisible properties operating in the world makes these feelings seem reasonable, and disgust is the negative consequence of violating our sacred values.

WHAT NEXT?

In this chapter we have looked at an emerging biological understanding of the world that depends on categorization based on outward appearances and inferred invisible properties. Our mind design seems set to look for patterns and deeper causal explanations for the different kinds of things we think exist in the living world. This process leads to spontaneous untaught concepts of essences, life energies and holistic connection. Many of these beliefs can also be found in ancient models of the natural world where hidden structures and mechanisms were thought to reflect a supernatural order in the universe.

While these intuitive concepts have true scientific validity to some extent, our naive way of thinking about them leads us to attribute additional properties that would be supernatural if true. For example, supernaturalism forms the basis of belief for those who advocate the sympathetic power of diluted potions and magical foods that share some resemblance to the affliction under question. In these situations, belief alone is sometimes sufficient to produce the desired result even though there is no active ingredient in the potion or food. Like the illusion of control discussed in chapter 1, believing that you will benefit is sometimes good enough.

Such beliefs also influence the way we see ourselves as members of a group. In particular, our supersense leads us to infer something essential and integral to the group that should not be violated or contaminated by outside influences. When this happens, we feel revulsion and disgust. These are emotional states triggered by mechanisms that exhibit many supernatural properties of sympathies, antipathies and spiritual contamination. In this way, our supersense operates to unite the group members by shared sacred values.

Groups are held together by these sacred values. All humans can be disgusted and we would be very suspicious of anyone who did not experience this particular emotional response. When someone says that they could easily wear a killer's cardigan, we identify them as an individual not prepared to share the group's sacred values even when these values are purely arbitrary. This is because our supersense makes these values seem reasonable because of the moral indignation we experience fuelled by our intuitive emotional system. As social animals, we depend on our supersense, even when it flies in the face of reason.

In the next chapter we examine how this supersense can lead to some very bizarre beliefs and practices where we think we can absorb someone else's essence.

CHAPTER SEVEN

WOULD YOU WILLINGLY RECEIVE A HEART TRANSPLANT FROM A MURDERER?

THE HUMAN BODY is made up of about two-thirds water. Maybe this explains our proclivity to describe other people with liquidized language, especially those with whom we may have to share some intimacy. Some people are slimy, while others are wet. Someone can be drippy, whereas another can just ooze charm. Is it a coincidence that these descriptions reflect comparisons with slime, a substance usually associated with disgust?

Like food, certain people can be yummy, whereas others can be revolting. And in the same way that essential reasoning influences how we feel about incorporating food into our bodies, the same goes for connecting with other people. When Granny 'just wants to eat you all up', not only is she comparing you to something delicious, she may want to absorb you!

When we reason about others, our judgements are coloured by our sense of essential connectedness. At one level, humans are tribal: we belong to one particular group and not another. But we also see ourselves as individuals willing to share certain levels of physical intimacy with the group and with specific significant others. Love, hatred, and disgust toward others are fuelled by gut responses that forge our strongest social relationships, and we intuitively think in an essential way about the nature of these connections.

We think like this because we need to justify our emotions in a tangible way. For example, in one study, adult subjects were told that they were to be given a vitamin shot to study the effects on visual tasks. In fact, some were given a shot of adrenaline, without their knowledge. Adrenaline is the naturally occurring hormone triggered during times of arousal. It makes you breathe faster, your heart races, and your palms sweat. What did the subjects make of their change in arousal? It all depended on the context. While they were in the room awaiting the fake visual test, they were asked to complete a mood questionnaire. At this point, a confederate of the experimenter who was pretending to be a genuine participant started acting either very happy or very irritated. The subjects who were unaware that their faster breathing, racing pulse, and sweaty palms had been caused by a drug reported feeling either angry or happy, depending on the state acted out by the confederate.[1] Do you remember the Numskulls from chapter 5? It was like the boss Numskull in the head office was receiving memos from all around the body telling him that something was up and that he had to send out a press release to explain why the body was feeling so aroused. Conscious experience was the spin doctor making sense of the messages.

In another study, an attractive female experimenter stopped and interviewed male subjects as they crossed a very narrow footbridge over a very deep ravine.[2] After the interview, she gave them her telephone number. The measure of interest was whether they called her later. Twice as many males whom she stopped in the middle of the bridge called her in comparison to males who had been interviewed at the side of the bridge. The explanation was as cunning as the finding: males who were interviewed in the middle of the high bridge were physiologically more aroused by the danger of the situation but misinterpreted this physical response as sexual attraction to the female interviewer. So our experience of emotions is a combination of bodily sensations and our attempts to interpret them. We try to make sense of our sensations.

When we encounter someone who triggers an emotional response, we apply the same interpretive processes. We may not be able to say exactly what it is that we either like or dislike about the person, but we have feelings about him or her. For example, have you ever felt uncomfortable in the presence of someone and not known exactly why? Maybe she stood too close to you, or maybe he shook your hand longer and harder than you expected. Or maybe the person touched your arm during the conversation. Physical contact can be either charming or repulsive. Why? I think the answer is that physical contact leads to the belief in potential contamination during social interaction. If the person is someone we are inclined toward, such as a potential mate or someone we respect, then the contact is welcomed. If it is someone we don't like, then physical contact can be revolting. Both responses operate on the basis of psychological essentialism even when we are not fully aware of this threat of contamination. By assuming some exchange of essence, we can justify our response in terms of contamination. For example, members of the lowest caste in the Indian system were known as the 'untouchables': they were deemed to be so disgusting that a higher-caste member would be contaminated by contact with them. Although the term 'untouchable' was officially abolished in 1950, it still operates today as members from different castes maintain various degrees of physical separation.[3] The same was true for the segregation that operated in the United States and the apartheid system of South Africa.

Calling people names such as 'filth' or 'vermin' not only dehumanizes them but also leads others to treat them as essentially different and contaminated. How else could a Hutu neighbour butcher a Tutsi child with a machete if not because the child had ceased to be human and become a cockroach?[4] Essentialism justifies whether we embrace others or shun them by providing a physical reason for our actions. Our actions may be socially motivated and for the good of the group, but they also feel right. Where do these feelings come from, and how do we link them to others?

I think the answer lies with children's developing essentialism, combined with a developing notion of spreading contamination. It is easy to see how such thinking can start to shape the way we respond to living things that we essentialize, most notably other humans. If essences are thought to be transferable, we will not consider ourselves isolated individuals but rather members of a tribe potentially joined to each other through beliefs in supernatural connectedness. We will see others in terms of the properties that make them essentially different from us. Such an idea suggests that some essential qualities are more likely to be transmitted than others. Youth, energy, beauty, temperament, strength, and even sexual preference are essential qualities that we attribute to others. Hence, we are more inclined to think that these qualities can be transmitted compared to, for example, hair colour, the ability to play chess, or political persuasion, which are more likely to be regarded as nonessential attributes of individuals that are more arbitrary and can change over time.

The more essential a quality is deemed to be, the greater the potential for contamination. Furthermore, as we have seen with the killer's cardigan, this reasoning is always biased to assume a greater potential for negative compared to positive contamination, possibly because, as we saw with respect to disgust in the last chapter, evolution is more geared towards protecting us from harm by making us sensitive to threat. Nevertheless, there is plenty of evidence that the supernatural belief that we can absorb the good essences of others is common throughout our culture, practices and attitudes.

DRACULA WAS A GIRL

Let's begin with a horror story. Horror stories often frighten us because they include abominations and violations of our intuitive essentialism. One of the most obvious examples in today's popular culture is the vampire myth. Vampires have existed in folklore for thousands of years and are found throughout the world's civilizations. Every culture has tales of the undead who seek vital essences from the living. Of all the various monsters created over the millennia, Bram Stoker's story of Count Dracula, published in 1897, is the most famous.

It is often thought that Dracula was loosely based on the sixteenth-century Romanian prince Vlad Dracula, known more charmingly by his nickname 'Vlad the Impaler'. Prince Vlad was particularly successful at defending Romania against the invading Turks and delighted at skewering his victims alive on sharpened wooden poles. However, it seems that Stoker took only the name for his character from the Romanian prince. The Irish author was no doubt more strongly influenced by events at Switzerland's Lake Geneva in 1816, when a bunch of Gothic writers, including Mary Shelley, spent an evening in the house of Lord Byron and Dr John William Polidori dreaming up horror stories to frighten each other. Shelley came up with Frankenstein, another abomination tale of essentialist violation, whereas Byron told a tale of a vampire that was later published by Dr Polidori under Byron's name. The creature described in Byron's 'The Vampyre' was unmistakably Lord Byron himself, depicted as a fated nobleman with piercing eyes. However, the historian Raymond McNally thinks that Stoker's Dracula was also strongly influenced by a woman, the sixteenth-century Transylvanian countess Erzsebet (Elisabeth) Báthory, who tortured and murdered 650 women and supposedly bathed in their blood to rejuvenate her beauty.[5] This is why Count Dracula was an ancient Hungarian nobleman who had a passion for blood and never seemed to age.

The Countess Báthory was one of the most beautiful and intelligent women in Romania, but also the most depraved. According to the legend, one day she violently struck one of her servant girls across the ear, causing her to bleed onto Elisabeth's hand. At first the countess was enraged, but she noted that as the blood dried her own skin seemed to take on the youthfulness of the younger woman. This was said to be the origin of her passion for bathing in the blood of young girls, who were trussed up, then had their throats slit and their bodies drained for the rejuvenating juice. At least the blood-lusting countess did have the courtesy to pay for the burials of her victims.

Eventually the body count mounted, and the local priest refused to bury any more of the girls from the castle who had died under suspicious circumstances. Undaunted, the countess and her servants gave up all pretense of secrecy and simply dumped the bodies in the neighbouring countryside. It was when four bodies were casually thrown over the castle walls in full sight of the locals that they eventually complained to the king.

When Romania's King Matthias II, who also happened to owe Elisabeth money, was alerted to the sadistic activities of the countess, he saw a perfect opportunity to kill two birds with one stone, as it were. On 29 December 1610, he ordered a raid mounted on her castle, where further bodies of young girls were found. The arresting officer was Elisabeth's own cousin, and in an effort to cover up the family scandal and save the countess, the four servants implicated in the murders were quickly tried and executed by being burned alive. One was mercifully spared the torments of the flame with a beheading. However, Countess Elisabeth Báthory never faced trial, and her cousin had her walled up in her castle, where she died three years later.

Countess Báthory was a sadistic killer, though it is doubtful that she actually took baths in her victims' blood. When the records of eyewitness evidence given at the trials in 1611 surfaced two hundred years later, there was no mention of bathing in blood. Certainly, the countess had been drenched in it. She was more of a cannibal than a vampire, as she had been seen to bite chunks of flesh from the young girls, including their breasts. Maybe the legend of bathing in blood for vanity was more acceptable than the possibility that the beautiful, intelligent countess was really a depraved, psychotic murderer.[6]

THE FOUNTAIN OF YOUTH

Bathing in blood to reduce the signs of ageing is just one of the folk myths that humans have generated in their search for eternal youth. Sometimes fact is stranger than fiction. As we grow old, we are increasingly concerned about how we are ageing, and most of us, given the opportunity, would prefer to look younger than older. One of the world's most valuable industries is rejuvenating cosmetics. This business is estimated to be around £6.4 billion in the United Kingdom alone. The average British woman will spend £186,000 on cosmetics in her lifetime, and most of this will be on rejuvenating creams.[7]

Almost all such cosmetic use is based on sympathetic magical beliefs. Various preparations are made from materials associated with vitality, such as the placenta or amniotic fluids. The infamous Tai Bao capsules of China are allegedly made from aborted human fetuses, though most capsules sold in traditional Chinese medicine are supposedly made with powdered human placenta. Whether human or animal, the claim of these rejuvenating products is that by applying ointments or swallowing capsules, you can halt, slow, or even reverse the signs of ageing. The fact of the matter is that few of these preparations have any active ingredients that can be absorbed through the skin. Moreover, our natural stomach acid easily destroys any such nutrients that we may swallow. Indeed, just like homeopathic medicines, many cosmetics have no active ingredients, thus avoiding the problem of satisfying the regulatory authorities. Still, the belief that the essence of youth can be imbibed is a very powerful one for most people.

In February 1998, the British viewing audience watched aghast when Channel 4 broadcast an episode of *TV Dinners*. In what is probably one of the most repugnant examples of exploitative TV, we saw the endearing celebrity chef Hugh Fearnley-Whittingstall devise a very special dinner for Rosie Clear to serve to her family and guests to celebrate the birth of her daughter, Indi-Mo Krebbs (no relation to the life-cycle guy). Fearnley-Whittingstall fried Mrs Clear's placenta and made a pâté to be served on focaccia bread. While her husband Lee had seventeen helpings, the dinner guests were less enthusiastic. Meanwhile, the viewing public was running either to their toilets or to their telephones. Channel 4 received a deluge of complaints and was severely reprimanded by the British Broadcasting Standards Commission over what was regarded as an episode that 'would have been disagreeable to many'. Why was the general public so upset? What was so wrong? Why were they morally dumbfounded? In an interview published some years later on his River Cottage website, Fearnley-Whittingstall identified the society's supersense as the culprit:

It was quite interesting to see people getting very hot under the collar without being able to put their finger on what it was exactly that they were upset about. It was the exploration of a food taboo and I think that's a very interesting area. No regrets, but it does get a little bit trying when people label you as 'the placenta guy', because there was a little bit more to it than that.[8]

Yes, there was a little bit more than that. Along with the placenta, the pâté featured shallots and garlic flambéed in red wine as well as a good sprig of supernaturalism.

THE TWIN WHO ABSORBED THE OTHER

It may seem beyond belief, but you really can absorb the physical essence of another person and incorporate them into your own body. In chapter 3, we briefly looked at the research on identical twins separated at birth and raised in different homes. Identical twins are spooky because they look like the same person, and this presents us with a problem. We naturally think of individuals as unique, in the same way we think of ourselves as unique individuals. That's what the word 'individual' means. But identical twins, who come from one embryo that split in two, seem like two copies of the same person. Remarkably, this process can sometimes go in reverse. When two people become one, we are again presented with the problem of what it means to be an individual. In the American version of Ricky Gervais's hit comedy *The Office*, the assistant to the regional manager is a character called Dwight Shrute. We are told that Dwight is a twin who absorbed his twin in the womb, and that gave him, as an adult, 'the strength of a grown man and a baby'. Dwight may be fictional, but his claim is not.

When Lydia Fairchild was called in by Washington State social services in 2002, she assumed it was just a routine interview for the welfare support she had requested since separating from her partner, Jamie Townsend.[9] The meeting turned out to be an interrogation and the beginning of a nightmare – truly something out of a Gothic horror story. Both Lydia and her partner were required to provide samples for DNA analysis to prove they were the parents of the children. When the results came back, Jamie was indeed the father, but Lydia was not the mother. At first Lydia thought that there must have been a mix-up, but she recalled a social worker saying to her, 'Nope. DNA is 100 per cent foolproof, and it doesn't lie.' The authorities treated her as a criminal. They suspected a scam. Fairchild, pregnant with her third child, faced prosecution for benefit fraud and child abduction despite the fact that there were hospital records to prove that she had given birth to her two children. Prosecutors called for her children to be taken away into care, and when she was due to deliver her third child, the court ordered that a witness be present. Fairchild's world was collapsing.

Luckily, someone else's nightmare would be her salvation. Four years earlier in Boston, fifty-two-year-old Karen Keegan received a letter with the results of blood tests that she hoped would be an answer to her prayers.[10] Karen was in need of a kidney transplant, and her family had undergone compatibility tests to see if any of them would make a suitable donor. Instead, she got quite a shock. The letter told her outright that two of her three sons could not be hers. They did not share her DNA and must have come from another woman. Suspicions were raised. Had there been a mix-up at the hospital? How could two of her sons have been swapped at birth? Karen knew she had given birth to all her boys. It is not something you forget easily or are likely to make up. Only after two years did doctors discover the answer. Karen was a chimera. The chimera is a mythological, fire-breathing, monstrous creature made up of the body of a lion and the body of a goat, fused together with a snake for a tail. However, in biology a chimera is an individual that hosts more than one source of unique DNA. How could this happen? The truth is stranger than any horror writer could imagine.

Early in her pregnancy, Karen's mother had twin embryos developing inside her. She would have given birth to twin daughters, but something changed, and the two became one. Karen had absorbed her twin sister. Karen possessed two sets of separate genetic code in her body. Biologically, she is two people. When they repeated the tests, they found the other set of DNA that matched that of her two boys. The results of this amazing case were published in the *New England Journal of Medicine* in

2002.[11] Luckily for Lydia Fairchild, when Karen Keegan's case came to light, prosecutors realized that they had made a terrible mistake. Further genetic tests were undertaken and, to Lydia's relief, she too was found to be chimeric. The case was dropped, but so was the request for support benefits. Lydia and Jamie got back together again soon after the nightmare.

Rare cases of individuals who are biologically two people challenge our view of what it means to be a unique individual. We think of them as two people because our concepts of unique persons, or males and females, require that two individuals occupy two separate bodies. They cannot occupy the same body. This would be unacceptable for a mind designed to categorize individuals. And yet these individuals have only one body and one mind. This is why we are so perplexed.

Similarly, hermaphrodites and mosaics challenge our fundamental understanding of what it is to be a human being. They may be rare, but they are not supernatural. They are simply natural variations that occur in the random genetic crap shoot of life. But our intuitive biology simply does not readily allow for such exceptions to the rule. We treat these individuals as freaks because they violate our natural order. If identical twins look alike, then they must be telepathic. If some unfortunate sufferer has a skin disorder that makes him look like an alligator or an elephant, maybe he also behaves that way.

Ironically, the same intuitive biology that leads us to confusion when categorizing individuals readily leads us to beliefs about individuals that would be supernatural if true. We may treat others as unique because they occupy separate bodies, but essentialism also leads us to think that individuals have essential properties in their bodies that we can absorb into our own. This is no more dramatic than in the cases where we literally incorporate another person into our own body.

THE STRANGE TALE OF ARMIN MEIWES

The idea that you can absorb someone's essence is a recurrent theme in explanations of cannibalism. However, cannibalism is a controversial topic among academics, who argue about whether it has really existed and why it may have been practiced.[12] The claim that it never existed seems undermined by research on the human prion disease Kuru, which is a variant of Creutzfeldt-Jakob disease, the human version of 'mad cow' disease.[13] Kuru was particularly common among the Fore tribe of Papua New Guinea, where the word 'kuru' means 'trembling with fear'. It is now thought that the disease was transmitted through the cannibalistic practice, up until the 1950s, of eating rather than burying dead relatives. Unfortunately, the most digestible but also most heavily contaminated portion of the deceased was the brain, which was prepared especially for women, who then easily transmitted the disease to their young children and babies. Women and children became the most vulnerable victims. Even though the cannibalistic practice was outlawed fifty years ago, the incubation period of Kuru is such that there were still new cases up until the 1990s, indicating that the disease had lain dormant in those children.[14]

The case for cannibalism is further strengthened by Richard Marlar, who has being going through the 'motions' of the ancient Puebloan Indians of the American Southwest known as the Anasazi. The motions are the poo that archaeologists found around campsites where the charred remains of human bones were found in cooking pots dating from around the twelfth century, which led to a controversy about whether the Anasazi had practiced cannibalism. This was resolved by biochemical analysis of the post-meal turds found at the campsite, which were shown to contain human protein. The only way that protein could have gotten there was if it had been eaten.[15]

So cannibalism was practised, but whether the idea was to absorb another's essence is less straightforward, as the reasons for the practice varied. It also depended on whether the consumed were enemies or relatives and on how much of them was eaten. The Wari tribe of South America would eat tribe members as a funerary ritual, whereas the Kukukukus tribe of Papua New Guinea preferred to eat their enemies but smoke their relatives.[16] When an enemy prisoner was captured, the men broke his legs with clubs so that he could not escape and then let the children play at stoning him to death. He was then chopped up, wrapped in bark, and cooked with vegetables in a traditional pit oven. If the victim was young, the muscular parts were given to the village boys to eat so they could absorb his power and valour. In contrast, deceased relatives were placed in their hut, where a fire was lit and the body was gradually smoked over the course of six weeks. In their belief system, the spirit was still present, and the survivors behaved accordingly, treating the leathery corpse as if it were still alive.

Such practices have long since disappeared, but every so often the taboo of cannibalism surfaces from the underbelly of human depravity. Following a tip-off about some weirdo posting ads on the Internet about wanting to eat men, the police raided the home of forty-two-year-old Armin Meiwes in the small German town of Rotenberg in 2002. What they found was truly grisly. Armin had a freezer containing human body parts and a video recording of the evening he killed and butchered his victim. That was just the beginning. The tale would develop into an even more shocking case of cannibalistic essentialism.[17]

A year earlier, Armin had advertised on an Internet chat room dedicated to sadomasochistic discussions, looking for a young man to slaughter and eat. Apparently, talking about cannibalistic fantasies is not that uncommon in Germany. Unbelievably, a forty-three-year-old Berlin engineer, Bernd Brandes, replied. In fact, Armin had half a dozen men come and visit him, but only Bernd apparently was willing to see it through to the end. Bernd harboured a real desire to be eaten. Following a brief e-mail exchange, they agreed to meet at Armin's home.

On the fateful evening of March 9 at Armin's house, Bernd Brandes swallowed twenty sleeping tablets downed with a half-bottle of schnapps. He then begged Armin to cut off his penis for both of them to eat. He wanted to be eaten alive! After an initial failed attempt with a blunt knife, Armin successfully cut it off. Bernd had difficulty eating his own manhood, as it was too chewy, and so Armin tried frying it with garlic but ended up burning the meal. Bleeding heavily, Bernd decided to take a bath. Meanwhile, Armin went downstairs to read a *Star Trek* novel. After a few hours, he returned upstairs to finish off Bernd with a kiss before stabbing him in the neck. He then butchered the body, putting the pieces in his freezer cabinet next to the frozen pizza. He buried the head in the garden. The whole incident was recorded on videotape, proving that Bernd Brandes had been not only a willing victim but also an encouraging one. By the time the police arrived in December 2002, Armin had eaten twenty kilograms of Bernd cooked in olive oil and garlic, washed down with South African red wine.

The media frenzy that followed brought up the obvious questions. Why did Armin do it? Armin claimed that from an early age he had wanted to eat another person. More disturbing to contemplate was how anyone could willingly want to be eaten. How could Bernd Brandes want such an horrific death or, for that matter, attempt to eat his own penis?

We can only speculate about Bernd's motives, and getting answers from Armin is proving to be difficult. I have made several requests to set up a meeting with Armin Meiwes, who is now serving a life sentence in Germany, but so far these requests have been declined. However, the available reports and testimony indicate that both men had a perverse sense of essentialism, vitalism, and holism.

In his e-mail reply to the initial posting, Bernd said that he wanted to exist inside another man's body. Clearly, he believed in an afterlife inside someone else. He was like the puppet mouse dead inside the alligator that children believed would still have a mental life. Armin held reciprocal supernatural beliefs about his victim. He wanted someone to live on inside him. During police interviews, Armin reported that Bernd tasted similar to pork, but that with every mouthful his memory of Bernd increased. He felt much better and more stable with Bernd inside him. He also reported that his English had improved. Bernd Brandes had spoken fluent English. In the most recent interview in 2007, Armin said that Bernd was still with him.[18]

CELLULAR MEMORIES

I may never get the opportunity to question Armin Meiwes about his supernatural beliefs, but I have spoken to the much more amiable and approachable Ian Gammons, who lives with his good wife, Lynda, in the small village of Weston in Lincolnshire, England. Lynda and Ian have been married for over thirty years and share an intimacy over and beyond what most couples can ever expect to achieve.[19]

In 2005 Ian was suffering from kidney failure when it was discovered that Lynda was a match and would make a suitable donor. She didn't even hesitate, and the life-saving operation was a success. About two months after the operation, Lynda and Ian were out shopping when something peculiar happened. Ian turned to Lynda and said, 'I am really enjoying this.'

Ian and Lynda have always been very close but very different in their interests. Ian is a typical male who hates shopping, gardening, cooking, and all the pursuits that Lynda loves. The idea of Ian enjoying shopping seemed too strange. Ian began to cook and to enjoy helping out in the garden. Previously, he would have simply heated up a frozen dinner. When Lynda raised the topic of getting a pet dog, Ian agreed despite having been a cat person all his life. And the similarities go beyond hobbies and tastes:

> My experiences are still developing. I am becoming more intuitive and I have a greater awareness. In particular, we share many more dreams. Last night Lynda woke up and said that she had a weird dream of a white house in a green field by the sea. I had exactly the same dream. Is it true that our DNA is mixing? Is that how it could possibly happen?

Ian is a soft-spoken man who genuinely wants to know how to explain his experiences. He is not your typical New Age hippie talking about essences, vital life energies, or the connectedness of the cosmos. The only sensible answer, he feels, is that he and Lynda now share a link because part of her is inside him. He has absorbed part of his wife and is turning into her in a small way.

It's not an uncommon report in transplant patients. Around one in three transplant patients believe that they inherit the psychological properties of the donor.[20] The most famous example was Claire Sylvia, who received the heart and lungs of a young man in the 1980s.[21]

Following the operation, she developed a taste for drinking beer and eating chicken nuggets. For a ballerina, this was a strange departure. More bizarrely, she found herself attracted to short blonde women. The deceased's girlfriend had been short and blond. And, yes, he liked beer and chicken nuggets, which were found in his coat following the motorcycle crash that killed him.

Such reports are claimed to be examples of cellular memories, a supernatural belief that the psychological aspects of an individual are stored in the organ tissue and can be transferred to the host recipient. Some claim that each cell of our body is connected. If the brain creates the mind and brain cells contain the psychological states of memory, then other cells in the body share this information. On the surface, something like Ian's belief that he had incorporated Lynda's mental states through her transplanted DNA does seem logical.

At one point, there seemed to be some scientific evidence for such a bizarre notion. James McConnell is a controversial figure in the science community. In the 1950s and 1960s, he was conducting experiments on simple worms to measure how long it took them to learn a maze.[22] Having trained a bunch of these flatworms to slither around the maze, he then did something very unusual. He chopped the trained worms up into small pieces and fed them to untrained worms. These cannibal worms now learned to slither around the maze faster compared to other worms that had not been fed the cannibal diet.

Further studies with rodents seemed to suggest that naive animals fed the bodies of trained animals learn to run mazes more quickly.[23] How could this be if it was not cellular memory? It turns out that the training involved stressing the animal with electric shocks so that it would avoid repeating mistakes in the maze. Remember John Watson and Little Albert and conditioning behaviour? This kind of stress releases hormones that stay in the body. It's one of the reasons slaughterhouses try to reduce the stress of livestock, because the changes associated with stress affect the quality of the meat. When the hearts and livers of trained mice were fed to novice mice, it produced a measurable difference in the latter's performance in learning to avoid shock. Was this evidence of memory transfer? No. If mice that had never been trained on the maze were simply stressed by being rolled around in a jar and then killed and fed to other novice mice, these novice mice also showed improved learning on the maze.[24] It was not a memory that was imbibed, but a hormonally enriched heart or liver. As happens when you pop a pep pill to study for a test, you learn much faster if you are more aroused. No reputable scientist does this kind of research today. Still, this has not stopped the spread of the cellular memory hypothesis, which can still be found in school science textbooks.

One has to question the logic that motivated James McConnell to do such a bizarre experiment, but clearly he felt that knowledge could be transferred by ingesting the body of another. Like many examples of pseudoscience, it is difficult to make the distinction here between natural and supernatural reasoning, since McConnell's hypothesis had surface credibility. Eating a trained animal made a difference on a memory task, so why not cellular memory? This line of research is now discredited by the scientific community but still cited as evidence for transplanted memories by those who believe in supernatural connectedness. Some case studies seem to stretch the bounds of credibility.[25] However inexplicable Ian and Lynda Gammons's experiences may seem, they do not seem beyond coincidence or reason. More difficult to explain away are cases such as the little eight-year-old girl who received the heart of a murdered ten-year-old. It was claimed that she started to experience terrifying nightmares and was eventually able to provide a detailed description of the man who killed her donor, enabling the police to capture and convict the murderer.

Such stories are myths that perpetuate supernatural beliefs. Relatives, patients, and those considering organ transplantation must be influenced by intuitive essentialism. This explains why there is a willingness to believe that we can inherit the psychological properties of another person through their organs. While it may be comforting to the families of donors to think that some essence of their loved one lives on, it may even have a negative effect when it comes to organ donation. Eternal essence may be a comforting notion to some relatives, but it may persuade others not to give consent in the belief that the relative still lives on in another. And what about recipients? How do they psychologically adjust to having someone else's organs inside them? In one case, a British teenager was forcibly given a heart transplant against her will because she feared that she would be 'different' with someone else's heart.[26] She was more frightened of losing her own unique identity than by the prospect of certain death. Such is the power of essentialist beliefs.

The Swedish researcher Margareta Sanner has been asking people what they think about organ transplantation and getting some very interesting responses.[27] She found that moral contagion was a major factor ('what if it comes from a sinful man?'), as were concerns about xenotransplantation – the substitution of animal organs for human ones. When offered a choice of different organs, adults typically responded, 'The liver and kidney from a pig is okay, but I would only accept a human heart', or, 'Everything is in the heart; I neither want to give it nor take it.' One participant even thought that 'I would perhaps look more piggish with a pig's kidney.'

We recently examined these sorts of beliefs in healthy students by asking them to rate the faces of twenty people for how attractive and how intelligent they looked and then for how happy the students thought they would be, if they were dying from cardiac failure, to receive a transplanted heart from each person.[28] Having initially rated the face of each potential donor on all these measures, we then told them that half the people in the pictures were convicted murderers and the other half worked as volunteers. They were then asked to repeat the ratings for attractiveness, intelligence, and willingness to receive the person's donated heart. Not surprisingly, though all the ratings for murderers dropped, the biggest effect was on participants' unwillingness to receive a heart transplant from a murderer. The participants may have thought that the evil of a murderer is a tangible property that can be stored and transferred in a simple pump of muscular tissue.

And what about bigotry and racism? In 1998 Northern General Hospital in Sheffield, South Yorkshire, was severely criticized for accepting the organs of a donor on the condition that they could only be transplanted into a white patient.[29] Following a similar case in which the family refused to allow the organs of a dead man to be transplanted into a nonwhite patient, the state of Florida passed a law banning such restrictions on organ donation.[30]

One of Sanner's most intriguing findings arose from her interviews with patients who had received a kidney transplant from a living donor compared to those who had received a kidney from a dead donor.[31] Unlike Ian and Lynda Gammons, the patients with an organ from a living donor were much less concerned about incorporating aspects of the donor's personality than were patients who had received a kidney from a dead donor. Maybe the recipients of living donors were better prepared (these operations are planned well in advance) and knew the donor was still alive and well and in full possession of his or her unique identity. But the recipients of an organ from a dead donor knew that the person was no longer around and wondered if part of that person lived on inside them.

Clearly, psychological essentialism influences the way we think: as a donor, we may continue to live on in another person's body or, as a recipient, we may be changed by having another person inside of us. Such supernaturalism can even be found in that most common preoccupation of human behaviour: sex.

ESSENTIAL SEX

If you are male and over forty, you will understand why one of the first movies that had an enduring impact on me was Roger Vadim's 1968 *Barbarella*.[32] The opening sequence of Jane Fonda's weightless striptease aroused strange feelings in most prepubescent boys like myself, but it was a sequence much later in the movie that left the biggest impression on me. On arriving on an evil planet, our heroine enters the palace of pleasures, where Amazonian women are sitting around on big cushions smoking from a giant hookah pipe. Inside the glass bowl swims a young man. The women are clearly high on the intoxicating smoke. When Barbarella asks what they are smoking, the answer is chilling. 'Essence of man' comes the reply. For a boy on the boundaries of sexual awareness, this was a terrifying revelation. Was sex all about having one's essence absorbed?

Sex with another person is layered with essential, vitalistic, and holistic beliefs. It may be triggered by hormonally induced urges (feeling horny), sensory stimulation (smells, tastes, and sights), functional drives (I need to make a baby), and even cultural pressures (go on, it's expected), but our thoughts during copulation and when we think about copulation are seeded with supernatural notions. Being at one. Soul mate. Achieving a sacred union. In what must be one of the most embarrassing moments for any member of the royal family, Prince Charles talked of reincarnating as his mistress's tampon in a secretly taped telephone conversation to his lover. The nation was disgusted by the revelations from the 'Camillagate' tapes, and it may have been said as a joke, but such notions really just reflect a lover's desire to be intimate and incorporated into the loved one's body. This is because lovers want to achieve both a spiritual and physical union.

Even where people do it has a spiritual consequence. Recently a man and woman were arrested in an Italian cathedral after parishioners heard groaning coming from the confessional box. When the authorities pulled back the curtain, they found a woman down on her knees, but not in repentance. She was performing a sex act on the man whose groaning was due to carnal pleasure rather than moral angst. The couple argued that, as atheists, having sex in a church was no different to any other place. However, the church thought that the act was so sacrilegious that a special ceremony would be necessary to purify the box.[33] The box had been contaminated by the act. This sounds remarkably similar to the 'Macbeth effect' we encountered in chapter 2 and the use of exorcism rituals to cleanse places polluted by evil.

If you hold essentialist views, it is easy to understand how you might regard sex as potentially contaminating, with either positive or negative essential qualities, depending on how you view the other person. This is why rape is not only a physical abuse but also a psychological violation that leaves the victim feeling 'dirty'. For many, sex outside of a partnership of two people, whether forced or complicit, is unacceptable because the essential integrity of our partner has been defiled. Consider how the various sex acts rank in order of their essential overtones. I don't need to spell them out, but the more physical the contact, the penetration, and the exchange of bodily fluids, the more essentialist our attitudes to the acts are. Climax achieved through nonphysical contact with another may be perverted (dirty phone calls and even virtual sex when it arrives), but it is not as essentially disturbing as actual physical penetration.

Also, why do we find sex among the elderly generally disgusting and yet older people themselves are often still sexually active? Our overall preference to have sex with younger partners may be an evolutionary drive to mate with healthier, longer-living potential partners, but the disgust we feel when thinking about old people having sex is derived from essentialism. Such ageist beliefs are not trivial. The urge to have sex with younger partners leads to exploitation. The older, stronger, and more dominant seek out the more vulnerable for sex. This is because in many cultures sex with children is deemed to be a way to regain youth and vitality.

And look at what we actually do down there in the genital region. How can anyone enjoy the pleasures of a recreation area that has a sewage outlet running through it? We can only do it if we find the other person sexy. Otherwise, with a stranger we do not find sexy, it becomes totally disgusting. Why does sex with one partner invoke lust and the other disgust? My suspicion is that such attitudes stem from a psychological perspective rooted in the essentialist notion of a need to make a profound connection with another by spreading essential seed.

This kind of sexual supernatural reasoning is potentially dangerous. According to official statistics, nearly sixty children under the age of fifteen were raped every day in South Africa throughout 2001.[34] The actual figure is thought to be much higher, since only one in thirty-five cases are reported to police. Various bodies monitoring the situation believe that the victims are increasingly younger. One explanation for this trend is the so-called 'virgin cure' myth, which extends to raping babies.[35] In 2000 South Africa's Medical Research Council reported that 'belief that having sex with a virgin can cleanse a man of HIV has wide currency in sub-Saharan Africa'. A survey of over five hundred automobile workers revealed that one in five thought the virgin cure was true. The origin of the myth is sympathetic magic, and it can be traced back as far as medieval Europe. However, I fear that the pandemic of HIV/AIDS is only going to lead to an increase in the occurrence of such attacks as desperate sufferers try by any means to cure themselves. This is because education can have little impact on traditional belief systems. Despite having one of the most intensive programs of health education in the world on the causes and prevention of HIV/AIDS, studies reveal that South Africans still endorse both biological and supernatural explanations for the cause of the illness. These two belief systems – natural and supernatural – are not viewed by participants as inconsistent with one another but rather as complementary causal explanations. For example, people know that a biological virus causes HIV, but they argue that witchcraft is responsible for one person contracting the virus and not another.[36]

THE WEAPON SALVE

Essentialism, vitalism, and sympathetic magic have a long history in medicine, For example, the medieval 'weapon salve' was a popular treatment for wounds of conflict.[37] This was the idea that acting on the weapon that had inflicted an injury could heal wounds. Here is a recipe for weapon salve from the renowned fifteenth-century Swiss alchemist Paracelsus:

> Take of moss growing on the head of a thief who has been hanged and left in the air; of real mummy; of human blood, still warm – of each one ounce; of human suet, two ounces; of linseed oil, turpentine, and Armenian bole – of each two drachms. Mix all well in a mortar, and keep the salve in an oblong, narrow urn.

Once this ointment was prepared, it was important to recover the original weapon and dip it in the ointment. In the meantime, the wound was to be cleaned regularly with fresh water and bandages each day after the removal of 'laudable pus'.

The logic of the weapon salve reveals a number of supernatural misconceptions. The weapon had a sympathetic connection with the wound by virtue of the fact that it had inflicted it. The various ingredients for the salve were chosen because they had sympathetic affinity with the healing process. Some ingredients may have been chosen because they were believed to counteract the negative aspects of infection by exerting antipathetic forces to cancel them out. The gruesome ingredients of the potion demonstrate essentialist thinking. The use of human tissue reflected the belief that it possesses essential forces that can affect the healing process. Particularly prized was the tissue from those who had died healthy and young; no one wanted rejuvenating fat and blood from either the ill or old. Hence, most recipes called for the use of those who had been executed, the younger and more virile the better, as the young had more life force in them than the sick and dying.

The weapon salve treatment did actually work, but not through any supernatural mechanism. Rather, simply cleaning the wound and replacing the bandages each day enabled the body to fight infection, which was the most common cause of death. However, those who practised the treatment believed that it worked for all the wrong reasons. A similar story would emerge in another extraordinary episode from the history of Western medicine.

THE GONAD DOCTORS

Apparently the idea came from his time working as an unqualified young doctor in a Kansas slaughterhouse, where he noted the sexual prowess of male goats. Dr John R. Brinkley, or 'the goat gonad doctor', reasoned that if one could transplant the gonads of billy goats into men whose libido was flagging, those parts that old age had rendered ineffective could be reinvigorated.[38] Brinkley's reasoning was pure essentialism and

FIG. 16: The goat gonad doctor, John R. Brinkley. © KANSAS STATE
HISTORICAL SOCIETY.

vitalism coupled with a naive understanding that gonads are related to
sexual function. The animal transplantation studies were originally
conceived as an early application of sympathetic essentialist reasoning
– like begets like. If male goats are horny, and your libido is dropping,
then put a bit of billy in your works.

His first patient was an elderly farmer who complained of a low
sex drive and was willing for Brinkley to try inserting a portion of
goat gonads into his scrotum. Most individuals would be appalled at
the notion of deliberately inserting animal tissue into their body, as
opposed to their stomach, but when it comes to sex and ageing, human
history is full of bizarre practices believed to enhance, improve, and
prolong the sexual experience. By all accounts, Brinkley's farmer not
only survived the operation but also enjoyed a renewed lease on sexual

life, fathering a son whom he decided to name, appropriately, Billy. John Brinkley's meteoric rise to fame and wealth had begun. He would go on to perform thousands of such operations at around $750 a pop, and he became one of the most successful quacks in twentieth-century America. For $5,000, which was a huge amount back then, Brinkley transplanted human gonads harvested from young prisoners on death row. Over his lifetime, he would own mansions, planes, boats, and radio stations and stand twice for the governorship of Kansas. He even wore a 'goatee' beard to fit with his medical procedure. Eventually, the American Medical Association, frustrated at the extent and success of his goat gonad transplants, ran Brinkley out of the country, and he would eventually lose his fortune trying to re-establish his career abroad.

The notion that animal sexual glands would work as an elixir of life had been around for some time. In nineteenth-century Paris, the ageing Harvard physiologist Charles-Édouard Brown-Séquard had been making claims of rejuvenation from injecting himself with crushed guinea-pig and puppy testicles. Ouch, it makes me wince just to type this. Probably the most famous gonad doctor of the time was the Russian-born physician Serge Voronoff. He injected himself with Brown-Séquard's liquidized pet bollocks, but with disappointing results. Voronoff then thought that perhaps the tissue should remain intact, so he perfected the transplantation or graft technique. Initially, he used the family jewels of young criminals and transplanted them straight into the sagging sacks of ageing millionaires who could afford the operation. When he ran out of obliging youthful crooks, he moved on to various monkeys and apes. World leaders, captains of industry, and ageing actors all paid handsomely for operations, and soon animal gonad grafting was taking place all over the Western world, except in England. The English, being strong pet lovers, had banned animal vivisection but deemed it perfectly acceptable to transplant another man's bollocks.

Unlike Brinkley in the United States, Voronoff enjoyed the accolades of his fellow doctors in Europe for a period of time. In July 1923, *The Times* reported that at a meeting of seven hundred leading surgeons at the International Congress of Surgeons in London, Voronoff was applauded for developing the rejuvenation operaton that would make him a fortune substantial enough to afford an entourage of servants and mistresses.[39] However, as with Brinkley, eventually the tide of support changed when it became clear that Voronoff's claims could not be substantiated.

Although Voronoff's reputation was eventually destroyed, aspects of his research were sound. The testes produce the steroid hormone testosterone, which is an essential mechanism for the masculinization of males. In the womb, testosterone turns girl babies into boy babies. Without it, all boys would turn out to be little girls. That's why we all have nipples. Over the course of the lifetime, testosterone plays a role in the so-called secondary sexual characteristics that appear around puberty with the change in the genitals, body mass, and hair. In old age, testosterone levels become depleted. Among other symptoms of old age, lowered testosterone can reduce the sexual libido, and so hormone replacement therapy is one controversial treatment for the so-called male menopause. It also forms part of the transitional female-to-male gender reassignment in women who want to be surgically transformed into men. However, in its modern use, synthetic manufactured hormones are used to avoid both the problem of rejection of animal tissues by the human immune system and the risk of transmitting animal disease into humans.

It was this risk that brought Voronoff out of his relative obscurity in 1999 when an article published in the science journal *Nature* theorized that his early gonad transplantations to rejuvenate the limp libidos of old wealthy men had inadvertently transmitted the deadly virus HIV from monkeys to man.[40] How ironic if true. Once again, the animals get their revenge on their superstitious tormentors.

Under normal circumstances, the cells from one animal cannot replace the cells of another. Even human-to-human transplantation requires compatibility and drugs to suppress the body's natural immune defense to reject foreign invasion. The fact that gonad injections and transplantations seemed to work was due to the placebo belief that they would work. Although the logic behind the gonad doctors' treatments was essentialist in nature, it would ultimately lead to the discovery of the underlying mechanism of the yet-unknown hormones.

When Voronoff observed the effects of castration on men and animals, he saw how the absence produced an imbalance. He simply reasoned that replacing what was missing in an old man would redress the problem. A naive conception based on the sympathetic laws of magic led to a scientific reality.

HOLY WATER

When Charles I, the British king, was beheaded on a cold January morning in 1649, it was reported that the crowd surged forward to dip handkerchiefs into the royal blood as it dripped from the scaffold.[41] If true, one possible explanation for this grisly reaction may have been the belief that royal blood had curative powers because kings and queens had a direct connection with God. Certainly, the 'royal touch' of a king or queen was thought to cure the skin disorder scrofula, a form of tuberculosis. Essential adoration of saints and kings continues to this day.

The most visited site in the Italian province of Umbria is the fortified medieval town of Assisi, home to the Basilica of St Francis of Assisi, where the remains of Italy's most famous saint, St Francis of Assisi, can be found. The tomb of this thirteenth-century saint was not discovered until 1818, which is surprising considering that these were the remains of the individual responsible for the formation of the Franciscan order of monks. The original tomb had been concealed by a fifteenth-century pope, but when the mortal remains were rediscovered following nineteenth-century excavations, they were moved to the underground crypt that pilgrims can now visit today. On the day I was there, the temperature was a searing ninety-five degrees Fahrenheit outside in the blazing Tuscan sun, so, despite the hundreds of visitors crammed into the basilica, it was a welcome relief to file slowly into the cool underground crypt and shuffle past and around the large stone sarcophagus protected by a lattice iron frame.

The numbers were such that one had to simply go with the silent majority. There was no turning back. Whenever a whisper emerged in the crowd, a disembodied voice from some unseen church authority reprimanded and commanded us with a stern '*Silenzio*'. We were expected to maintain a reverential state. However, just as museums tell us, 'Please do not touch', it was understandable why visitors wanted to poke their hands through the iron grid to make physical contact with the ancient stone monument behind. Some were engaged in silent prayer as they touched the stone.

It was then that I witnessed something quite disturbing and essentialist in nature. A monk came in and watered the permanent flower arrangements at the front of the tomb. The water from the flowers started to trickle over the ancient stone. What I did not expect, and could not photograph because of the restrictions, was a sudden frenzy in those nearest to this part of the tomb. As if they had been parched beyond thirst by a desert sun, they pressed their faces against the grid trying to lick the water as it dribbled over the Holy Shrine. Fingers wetted by the excess water were licked in an effort to imbibe some of the vital fluid. Water that was probably drawn from an ordinary tap from the municipal supply only minutes earlier had become sacred by contact with the tomb. It was all too bizarre. Admiration and adoration had become essential contamination of ordinary water.

SACRED SOIL

Such essential thinking is at the heart of a business dream of Alan Jenkins and Pat Burke.[42] I met them last year on a Dublin chat show, where they were talking about their new business venture, the Auld Sod Export Company. Alan is an elderly, more reserved man and maybe a little too serious, whereas Pat is a much younger, jovial agricultural scientist who enthused about the new product they were selling in the United States: Irish dirt. Not just any old dirt, but true, authentic Irish soil. Alan got the idea when he attended a funeral in Florida and overheard the relatives lamenting that the departed could not be buried in the auld country and that just a little sprinkle of Irish dirt on the coffin would have been a comfort.

Dirt is full of microbes and potential contaminants. US customs and agricultural import restrictions are some of the most stringent in the world, and so the dirt would have to be thoroughly sterilized to remove any potential biohazard. This was Pat's role in the business – to produce the cleanest dirt in the world. Good enough to eat. It sells at $15 for a twelve-ounce bag, and one elderly New Yorker originally from Galway has ordered $100,000 worth so that he can have his Irish grave in Manhattan. The company is now branching out into shamrocks that can be grown in the United States in Irish dirt in time for the ever-popular St Patrick's Day celebrations. The belief is that somewhere in what must be the most sterile soil on earth the essence of Ireland remains. With such potential for psychological essentialism at work in the large expatriate Irish American community. Alan and Pat may have struck pay dirt.

During the Second World War, Germany invaded Yugoslavia, and the royal family fled to exile in London. King Peter II, the last king of Yugoslavia, married Princess Alexandra of Greece in 1944, and they were expecting their first son the following year. Anxious about the heir to the throne not being born in his homeland, King Peter II made a special request to Winston Churchill. For a single day in the summer of 1945, the British prime minister, Sir Winston Churchill, conceded room 212 of Claridge's Hotel in Brook Street, London, over to Yugoslavia so that Prince Alexander could be born in Yugoslav territory. A pot of Serbian soil was placed under his bed to add the essential ingredient to a political decision.[43]

And how did our vampire from the beginning of this chapter move around and remain safe during the daylight hours? By travelling in coffins that contained the dirt from his native Transylvania, of course.

WHAT NEXT?

In this chapter we examined ways in which humans can experience or seek out an intimate connection with significant others supported by beliefs that they can absorb another person's qualities. This experience can be either positive or negative depending on the properties that we believe we may incorporate. While biological contamination through viruses and microbial infection is a real mode of transference between individuals, we also believe that other non-physical properties such as vitality, morality, and even identity can similarly be transferred as if they were physical entities. Such beliefs may be based on a naturally developing notion of essences we infer when thinking about other individuals. I think these beliefs are a natural product of the way that we think about other people.

Essential reasoning comes from the gut as much as it comes from the mind. That's because it's based on intuitive feelings that stir the emotions.

Emotions are the fuel that fires the decisions we make. Without emotion, our decisions are cold and without feeling. This may be fine when choosing which newspaper to buy or socks to wear, but when it comes down to decisions about other people, emotions are important guides to how we think. If these people are significant others in our lives with whom we share some degree of interpersonal commitment, then emotions are essential – in that the relationship must have some emotional component to be significant and in that it is easier to understand the experience of emotion as coming from some inner truth about the person with whom we feel connected.

If our emotions towards others are based on essentialist reasoning, we should be able to demonstrate that the principles of essential contamination apply as well. Personal possessions, items of clothing, and former dwellings of significant others will take on something of the previous owner. In other words, we will start to treat inanimate things and objects as if they are tainted by the essence of significant others towards whom we hold some emotional stance. To do so, we have to see that other person as a unique individual.

CHAPTER EIGHT

WHY DO TRAVELLING SALESMEN SLEEP WITH TEDDY BEARS?

WHEN I LEARNED that *SuperSense* was to be published, one of the people I wanted to tell was Steve Bransgrove. Four years ago, I had wandered into Steve's tiny shop on a cobbled street in the ancient nearby Somerset market town of Frome. Steve Vee Bransgrove Collectables was an Aladdin's cave of memorabilia with items from bygone times such as postcards, tin toys, comics, medicine boxes, and all manner of common objects of no obvious worth. But people would pay good money for them, toys in particular. The objects were so evocative. If you closed your eyes, you could smell the decades pass you by. Literally, the shop had a wonderful aroma of the past, laced with the scent of Steve's hand-rolled tobacco.

I remember the day I became addicted. I had casually flicked through some picture postcards in a box and discovered one of Tommy 'Twinkle Toes' Jacobsen, the armless pianist. The publicity shot showed a jovial moustached man wearing a black tuxedo carefully balanced at a piano playing it with his bare feet! I was amazed that there was once a time when individuals like Twinkle Toes were considered celebrities. I bought the card, and that was the beginning of my brief collecting obsession. Over the next couple of years, I would visit Steve's shop regularly. At first it was postcards from vaudeville and freak shows. Then for some reason I expanded into black-and-white postcards of beautiful 1930s' movie starlets. Often on my visits to Steve's shop I had no intention to buy, but we would chat about collecting and the people (mostly men in his experience) who follow this strange pastime. On each visit I invariably left with yet another small addition to my collection.

Steve had many wonderful tales of the obsessive collector – the wild look in the eyes, the change in expression when some coveted item was discovered, the agitated voice. He used to keep items under the counter for his regulars in the full knowledge that they would buy what he had to offer. Steve remembered each customer's particular fetish. Like a drug pusher, he fully understood the power of the addiction, as both he and his wife Shirl were collectors too. Steve barely made a living out of the business, but he enjoyed it so much that I bet he would have worked for rent money alone.

Why do people do it? Collecting seems such an odd behaviour in a world of instant upgrades, duplication, and modern innovation. Why look backward? When I entered the collector's domain, I discovered a mirror world populated by legions of people who traipse around car trunk sales and flea markets every weekend seeking authenticity. Come rain or shine, these people were out in droves, looking for the original.

There is money to be made from collecting, but that's not the only reason people do it. Money simply justifies the urge in most. The actor Tom Hanks, wealthy by anyone's standards, collects pre–Second World War typewriters. He sometimes spends more money repairing them than they are worth.[1] Any collector can relate to this. For example, vintage cars are the folly of the extremely wealthy. It does not make financial sense to own such a collection.

Other people collect because of the joy of the pursuit of the missing piece. Such collectors are motivated to complete the whole set even if they cannot physically own the set. For example, trainspotters are people who collect train numbers. These individuals (mostly men) stand at the end of platforms of busy stations writing down the serial numbers of the different trains as they come and go. They are like bird-watchers, or 'twitchers' – the obsessed individuals (again usually men) who race up and down the country in an effort to spot as many different species of birds as they can find. This male passion for completing a set fits with Simon Baron-Cohen's theory that we mentioned in chapter 5 about men being naturally inclined to order and systems.

However, collecting to completion is only one part of the obsession. Many collectors are motivated by the emotion generated by objects and the connection that objects make with the past. Collectors relish the sentimental feeling one can get from having and holding something from another time. If the object is associated with a significant person or event, the sense of connectedness is heightened. We recently conducted a large study of adults' attitudes toward objects and found that, not only do we value authentic objects, but we also want to touch them.[2] That's why people will pay excessive amounts for Jackie Kennedy Onassis's faux pearl necklace or bits of Princess Diana's wedding dress. These authentic objects command distorted values in the mind of the collector.

Examples like these demonstrate that the urge to collect memorabilia can seem weird or strange, but Steve's theory was that people collect memorabilia that reminds them of their own childhood or of better times when they thought they were happy. Objects are tangible, physical links with the past that can instantly transport us back to earlier days through a sense of connectedness. People don't collect objects that make them feel sad. I am not sure what my motives were for accumulating postcards of side-show freaks and Hollywood starlets, but I readily appreciated the pleasure in discovering a comic annual or toy in Steve's shop that I had seen as a boy and the way it took me back over the years. Each object was like unexpectedly meeting a long-lost friend.

When I told him I was working on a book about child development and the origins of irrational behaviour, Steve had promised to share tales of his more famous clients and their guilty collecting secrets. If I got a publisher, I would be back to discuss this more, as there are few things more irrational than the human obsession for collecting.

As I approached Steve's shop to tell him the good news about the book deal, the first thing I noticed was that he was not standing in the doorway chatting to passers-by with his trademark coffee mug and rolled-up cigarette. I then saw the note taped to the inside of the window. My heart sank. Had he gone out of business? Surely not, as

I knew Steve ran the shop for the love of dealing in memories, not to make money.

The truth was worse. Steve had died prematurely only weeks earlier in a sudden and rapid decline, before I even got a chance to know he was ill. In the letter taped to the window, his wife Shirl thanked everyone for all the words of kindness, but she could not continue the business without Steve and the shop would close. I returned only recently to see that the tiny premises were cleared out entirely, leaving just a shell, with the note still stuck to the window. I was surprised to see how large the shop had really been; Steve had packed it with so many objects that it had felt cozy and cluttered in a comforting way.

FIG. 17: Steve Vee Bransgrove Collectables in Frome (2007), where I spent many a happy hour. AUTHOR'S COLLECTION.

It was like the guts had been ripped out of some big, friendly, furry animal. A bit like the man himself. I am sure such a sight would have broken Steve's heart.

For me, the most poignant aspect of this story was not so much the loss of Steve (we all have to go) but the realization that many of us agonize and fret about possessions when we are alive. We accumulate objects over a lifetime in the belief that objects are important. We covet simple inanimate things. We invest emotion, effort, and time, and to what end or purpose? Only the very major collections survive intact, and they usually include recognized works of art with a commercial value. These are not the things that most of us could ever own. Personal possessions are often of little financial worth, and yet during our lives we are often annoyed or upset if they are damaged or lost. That's because objects define who we think we are. We treat objects as an extension of ourselves. When someone dies, most of their possessions are distributed, sold, or handed down, but often they end up in the flea market or in the trash. It's sobering to see how pointless a lifetime of collecting objects seems once the collector is gone. Sometimes when objects become symbols for a significant other, however, they can take on essential value.

Michel Levi-Leleu last saw his father Pierre in 1943 carrying a cardboard suitcase when he left the safety of a refuge in Avignon, France, looking for a new home for his Jewish family. Michel never saw his father again, but sixty years later the suitcase would reappear at the centre of a legal battle over ownership.[3]

It was a terrible time when Michel's father and suitcase went missing. The Jewish Holocaust of the Second World War was one of the most atrocious crimes against humanity in modern times. For the half-million annual visitors today, one of the most disturbing displays in the museum at Auschwitz is the pile of battered suitcases that once contained all the worldly possessions of families who would end their days in the death camp. Each case was labelled with the name of the owner in the belief that they would be reunited with their belongings again. The Nazis knew that to maintain the charade people had to think that their possessions were going to be kept safely and returned to them at some later date.

In 2005 Michel visited the Shoah Memorial in Paris, which was hosting a temporary Holocaust exhibit that included some of the suitcases on loan from Auschwitz. He knew his father had died during the war, but he could not believe his eyes when he spotted the suitcase with the handwritten label reading PIERRE LEVI. He asked for it to be returned. When the Auschwitz museum refused to hand over the suitcase, Michel took the museum to court. In court papers the museum stated, 'The suitcases of prisoners deported to Auschwitz that are exhibited at the museum are among the most valuable objects that we have.' The Polish government backed the museum.

Museums thrive on displaying authentic items, but today many face legal battles for the return of items to the descendants or countries from which they were taken. For example, Britain has been locked in a diplomatic quarrel for some decades now to return the Elgin Marbles from the British Museum to Greece. In the United States, Native American tribes have demanded the return of sacred objects.[4] Many museums now display copies and replicas without telling the public, or at least they give the impression that what you are viewing is authentic. This is because people want to make the connection with the original item. But, like beauty, authenticity is often in the mind of the beholder.

Once again, this kind of reasoning is something I have experienced myself. The family expedition to the Niaux caves that I described in chapter 3 was not the first time I had visited a prehistoric cave. On a driving tour around France in 1990 I chanced upon the more famous prehistoric Lascaux caves in the Dordogne region.[5] It was an unexpected opportunity, one not to be missed. At the time I was not particularly knowledgeable about or interested in prehistoric cave paintings and equally did not understand French particularly well, but I had heard of the Lascaux caves, and they were amazing. The animal drawings, all carefully lit in a remarkably accessible underground journey, were breathtaking. I was so naive that I did not realize my error. It was only when I left that I picked up a brochure explaining that the cave I had visited was in fact a reproduction of the original cave nearby that had been closed to the public since 1963 because of the problem of corrosive breath on the original paintings. I felt stupid and cheated. If I had known, I probably would not have bothered with the tour. Thankfully, the trip to the genuine Niaux cave fifteen years later, where we stumbled around in pitch-darkness, restored my sense of wonder in prehistoric art. No matter how good a reproduction is, knowing that it is not original destroys any sense of connectedness such an experience generates.

ESSENTIAL ART

In 2005 Sotheby's in London sold *Lady Seated at a Vestral* for £16 million, following ten years of dispute about whether it was an original Vermeer masterpiece or a twentieth-century forgery attributed to the expert forger Han van Meegeren.[6] After it was announced that the picture was an original Vermeer, its value soared. Nothing about the picture had changed – only the expert opinion about who had painted it. This proves that the appreciation of art is more than how something looks. It also depends on who you think created it. Auction houses typically charge up to 20 per cent commission on sales, so it's no surprise that the Vermeer authentication was provided by, of course, Sotheby's own experts.

Provenance in collecting is the proof of originality. Collectors seek authentic originals with provenance because they are more valuable. But why are originals more valuable than an identical copy? One could argue that forgeries or identical copies reduce the value of originals because they compromise the market forces of supply and demand. In the same way that a prolific artist who floods the market with work undermines the value attributed to each piece, rarity means limited supply. For many collectors, however, possessing an original object fulfills a deeper need to connect with the previous owner or the person who made the item. I think that an art forgery is unacceptable because it does not generate the psychological essentialist view that something of the artist is literally in the work.

Such psychological essentialism has been taken to its logical conclusion in the world of contemporary art. This is especially true for the Young British Art movement of the 1990s. For example, one of the most notorious essentialist artworks is Tracy Emin's piece *My Bed*, which was short-listed for the 1999 Turner Art Prize and sold to the collector Charles Saatchi for £150,000. The piece was simply the artist's unmade bed surrounded by her soiled underwear, a vodka bottle, and crumpled cigarette packets taken from a time she spent several days in the bed owing to a suicidal depression. Other artists, such as living icons Gilbert & George, are equally notorious for works of art made from their bodily fluids and excrement. However, probably the most essential artwork is one that was regarded as a signature piece of the Young British Art movement. Marc Quinn's 1991 *Self* is a self-portrait sculpture of his head made from at least eight pints of the artist's own frozen blood transfused over five months. Saatchi bought *Self* for £13,000. Interest in the piece was fuelled by press reports in 2002 that workmen renovating Saatchi's kitchen accidentally unplugged the freezer containing the head.[7] However, *Self* was on display in the Saatchi gallery a year later, raising questions of authenticity. Because of the deteriorating nature of the material, Quinn remakes the sculpture every five years with his own blood. Saatchi sold *Self* in 2005 to an American collector for £1.5 million. One wonders what will happen to this work of art once the source of original material runs dry. Will Quinn's descendants be expected to replenish the supply of blood after the artist has died?

We all treasure sentimental objects from within our own lifetime that do not necessarily have an intrinsic worth other than their connection with a family member or a loved one. These objects are essentially irreplaceable. For example, engagement or wedding rings are typical sentimental items that are unique. If lost or stolen, most people would not regard an identical replacement ring as a satisfactory substitute, because these objects are imbued with an essential quality. Psychologically, we treat them as if there were some invisible property in them that makes them what they are.

But what if it were possible to make identical copies? Imagine that a machine existed that could duplicate matter down to the subatomic level, such that no scientific instrument could measure or tell the

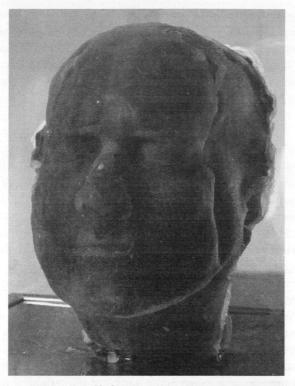

FIG. 18: Marc Quinn's *Self.* © **Marc Quinn.** PHOTOGRAPH BY STEPHEN
WHITE, COURTESY OF JAY JOPLING/WHITE CUBE GALLERY (LONDON)

difference between the original object and the duplicate – like a
photocopier for objects. If the object was one of sentimental value,
would you willingly accept the second object as a suitable replacement?
For most people, the answer is a simple no. Consider your wedding
ring. Let's assume that you are happily married and cherish the ring
of gold on your finger. Would you accept an identical duplicate even
though you could not tell the two apart? If you feel emotional, the
answer is most likely not.

Identical replacements are not acceptable because psychologically
we believe that individual objects cannot be replicated exactly even by

a hypothetical perfect copying machine. This attitude is based on the assumption that originality is somehow encoded in the physical structure of matter. We intuitively sense that certain objects are unique because of their intangible essence. However, such a notion is supernatural. Let me explain why with a much bigger example: a whole ship.

THE SHIP OF THESEUS

Early in the hours of a Monday morning in May 2007, a fire blazed through the nineteenth-century clipper the *Cutty Sark*, one of London's major tourist attractions docked at Greenwich. Initial reports from the fire crews at the scene indicated that almost all of the ship had been destroyed. However, the ship was undergoing a £25 million restoration, and Chris Livett of the Cutty Sark Trust confirmed that half of the ship's fabric had already been removed. He said that the ship had survived many potential disasters in the past and that the current crisis would be overcome.[8] Even if the *Cutty Sark* can be restored, questions remain: Will it still be the original? When does restoration and repair become replacement? How much of the original can be replaced before it is no longer regarded as the same thing?

Whether it is a ship or a decaying work of art at issue, such questions about restoration and conservation raise the philosophical problem of identity. If the material fabric of an object is replaced in its entirety, can the resulting object ever be said to be the original? What proportion of replacement is acceptable before the object ceases to be the original? What if the renovation is gradual?

Such issues raise interesting questions about how the mind represents objects in terms of originality after they have been repaired. The custodians of the *Cutty Sark* were quick to point out in early press releases hours after the fire that at least half the ship was already safely in storage. How did they come up with such a proportion so quickly? Was it based on weight or volume? I suspect it was based on the intuition that sudden damage to more than 50 per cent of the ship would have been regarded as the catastrophic loss of the original.

This modern act of vandalism reminds us of Plutarch, the Greek historians who told of an ancient conservation project undertaken to preserve the ship belonging to the legendary Athenian king Theseus. Over the years the boat was kept in service by simply replacing the timbers that wore out or rotted with new planks, to the extent that it was unclear how much of the original ship remained. Plutarch asked whether this was still the same ship. What if the replaced planks had been kept and reassembled to form a second ship? Which ship, Plutarch asked, would be the original Ship of Theseus?

Psychologists have begun to look at these questions of authenticity and essential reasoning towards objects in the lab. For example, five- and seven-year-olds and adults were shown a picture of Sam's 'quiggle', a nonsense object created for the purpose of the study.[9] One group was told that it was an inanimate paperweight, and the other group was told that it was a type of weird pet. Participants were then told that Sam went away for a very long holiday and that while he was away various parts of the quiggle were gradually replaced. The participants were presented with a sequence of photographs showing how the quiggle changed each week. Finally, they were presented with two pictures: one of the quiggle that had been gradually transformed and now looked completely different from the first picture, and another of the quiggle made out of all the removed parts recombined to look like the original quiggle before Sam left. The question of interest was, after he returned from his journey, which was Sam's quiggle?

Children and adults were more likely to say that the gradually transformed quiggle was the original, even though it looked very different and the reconstituted quiggle made of the replaced pieces was more similar to the picture of the original quiggle. This effect of continued identity over change was strongest when the quiggle was thought to be a type of living animal. This response fits with the intuitive biology of young children we discussed earlier. They understand that living things have something inside them that makes them what they are and that, despite outward appearances and changes, living things are essentially the same. This way of thinking is perfectly reasonable because we as individuals undergo significant change over our lifetimes as we age.

Not only does our outward appearance change radically, but so do our insides. The body is continually replenishing its own structures and cells over the course of a lifetime, though few of us are aware of such biological details. For example, if you are in your middle age, most of your body is just ten years old or less.[10] Now that's a fact worth remembering when we consider our attitudes toward ageing bodies!

However, for the older children and adults, even the quiggle that was described as a paperweight was regarded as the same object after undergoing radical transformation so that it no longer looked anything like the original. Younger children did not make this judgement. These findings show that with age we increasingly think of an object as being the same even though all of it is replaced with entirely new parts. In other words, there is something in addition to the physical structure of an object that makes it what it really is. What is this additional property? Where is it? It does not really exist, but we infer that it must be there. This is the essence that defines an object. As we grow older, we increasingly apply our developing intuitive essentialism to significant objects and living things in the world. I think this psychological essentialism is one of the main foundations of the universal supernatural belief that there is something more to reality. Where and when does this inclination to treat certain objects as special and irreplaceable first emerge? Remarkably, it may begin as early as in the crib.

SECURITY BLANKETS

I was listening to the radio this morning when I heard Fergie's latest hit record, 'Big Girls Don't Cry'.[11] In the chorus, she sings, 'And I'm gonna miss you like a child misses their blanket.' Any parent who has raised a child attached to a blanket or teddy bear will readily know what Fergie is singing about and will be familiar with the intensity of emotion that such a loss can incur.

Estimates vary, but somewhere between half and three-quarters of children form an emotional bond to a specific soft toy or blanket during the second year of life. These items have various names, including security blankets, attachment toys, and transitional objects. They are 'security blankets' because children need them for reassurance when frightened or lonely. They are 'attachment items' because of the emotional connection the child forms with them. And they have been called 'transitional objects' because one theory is that they enable the infant to make the transition from sleeping with the mother to sleeping alone. This may explain why such objects are more common in Western culture whereas they are relatively rare in societies such as Japan,[12] where children continue to sleep with their mothers well into late childhood.

Although I was familiar with security blankets from the Linus character in the *Peanuts* comic strip, who is always seen carrying his blanket around with him, I did not fully appreciate the significance of such behaviour until my first daughter developed an excessive attachment

FIG. 19 A desperate 'Wanted' poster to return Laurel's 'Mouse', lost in a Bristol park. IMAGE © KATY DONNELLY.

to her 'Blankie', a multicoloured, fleecy blanket that was in her crib. Blankie went everywhere with her. If she became upset, she needed to have Blankie.

It can be disastrous when these items are accidentally lost. When I was talking about the items on radio phone-ins, I fielded calls from distraught parents who had suffered the consequences of their child losing their attachment object. It's a fairly common tragedy and, as with lost pets, parents will post missing notices, such as the one shown in the picture overleaf.

I contacted the mother of the little girl who posted this note in a local park. I was curious to discover whether her little girl's mouse was ever found. She told me that it hadn't been found, but, remarkably, someone saw the plea and took the picture of the missing toy to their grandmother, who knitted a copy of 'Mouse' using the same materials. Despite the kindness of strangers, little Laurel did not accept the replacement mouse. It did not have the essence of the original.

Around the time children begin school, most abandon their attachment objects. Still, many children grow into adults keeping their prized possession. When I began researching this phenomenon, I surveyed two hundred university students and found that three-quarters said that they had had a childhood attachment object, usually a stuffed toy or blanket. There was no difference between males and females in remembering that they had had such objects. However, most males had abandoned their attachment objects by around five years of age. In contrast, one in three female students still had their childhood object as an adult. These figures are based on a straw poll of memories from a select group of students and cannot be used to describe the general population. Most people are too embarrassed to admit that they still have their sentimental childhood objects. However, a recent survey of two thousand solitary travellers by the UK Travelodge hotel chain revealed that one in five men slept with a teddy bear – more than the female travellers.[13]

Attachment to objects may be formed in childhood, but it's a behaviour that knows no age limit. Pamela Young is eighty-seven years old. On reading of my research into attachment objects, her son, Rabbi

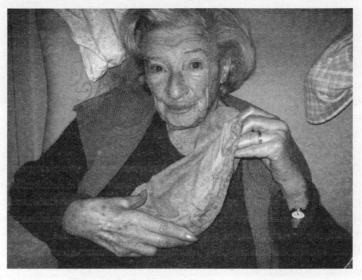

FIG. 20: Pamela Young with 'Billy' in 2007. IMAGE © RODERICK YOUNG.

Roderick Young, got in contact to tell me about the most important possession in her life, a pillowcase from her childhood crib she calls 'Billy'.

Pamela has had Billy for as long as she can remember. She sleeps every night with her head on Billy, clutching him with her right hand close to her face. Pamela has only ever been separated once from Billy – during an air raid in the London Blitz of 1944. She was staying in the Savoy Hotel with her first husband when the sirens sounded for guests to take refuge in the air raid shelters below. When she discovered that she had left Billy in her room, Pamela had to be physically restrained from returning to her room. Such is the power of sentimental objects. Roderick tells me that Pamela has requested that Billy be placed in the coffin with her, a promise Roderick intends to keep.

THE COPY BOX MACHINE

What is it about attachment objects that children cherish? Obviously, the physical properties are critically important for identification, but Paul Bloom and I suspected that the attachment ran much deeper than just the smell, sight, and feel of these objects. Why are they so irreplaceable? We decided to build the copy box machine to answer this question.

According to various physicists, duplicating machines are theoretically possible, just unbelievably improbable because they require vast amounts of energy and memory.[14] Undaunted, we built a copying machine on a shoestring budget. It comprised two scientific-looking boxes with knobs and dials with flashing lights.[15] Each box opened from the front so that an object could be placed inside. We showed this 'machine' to four- to five-year-olds and demonstrated how it worked. We placed various toys in one box, activated the machine, stood back, and waited for several seconds. After a moment, the second box activated by itself to alert the operator that the copy had been made. It was amazing. When both boxes were opened, there was a toy in each that looked exactly the same. We copied various toys, making exact duplicates of the original. The children were convinced that the machine actually worked and did not figure out that there was a second hidden experimenter behind the machine feeding in duplicate objects. The critical test was whether or not children would allow us to copy their own toys. Of course, we could not really copy their objects because we did not have duplicate blankets and soft toys. They simply had to decide which box to open in order to retrieve an item.

We identified two groups of children: those with favourite toys but no particular attachment to them, according to their parents, and those children who needed to sleep with the object every night. Children with favourite toys thought the machine was 'so cool' and happily offered their toys for duplication and even preferred to choose the box that was thought to contain the copy. In fact, they were often disappointed when we opened both boxes and confessed that the whole thing had been a trick. In contrast, the children with attachment objects either would not allow us to put their item in the machine in the first place or emphatically demanded return of the original. When we explained the setup, they were relieved to discover that their object could not be copied. Children did not want an identical copy of their attachment object. I think that they wanted the original back because a copy would lack the essential unique quality that we imbue sentimental objects with.

What about objects that did not belong to the child? Could we find evidence that they also thought others had unique possessions? Would they also treat the original and duplicates as essentially different? The copy machine was put into service again to look at the origins of authenticity and the value we put on memorabilia. We showed six- to seven-year-olds a metal spoon and a metal goblet and told them that one item was special because it was made of silver and the other was special because it once belonged to Queen Elizabeth II. This time it was easy to produce an identical copy, since we had bought two of each in advance. When we produced the second copy, we asked the children to value each item with counters. If the object had been described as special because it was made of silver, the children placed equal value on the original and the copy. It was made of the same stuff. However, if the item was said to have once belonged to the Queen, the same children valued the original over the copy. Something in the original could not be duplicated. Was it simply an association, or did children think that there was something of the previous owner in the object?

HOW TO SPOT A DALAI LAMA

The essential nature of objects is central to Tibetan Buddhist beliefs. According to eyewitness testimony from 1941, the bureaucrats in charge of finding the reincarnated Dalai Lama tested a two-year-old boy, Tenzin Gyatso in the remote village of Takster in the Qinghai province of China. Their technique was simple. Tenzin was presented with the personal possessions of the previous Dalai Lama, the 13th, along with a selection of inauthentic similar or identical looking items. When presented with an authentic black rosary and a copy of one, Tenzin grabbed the real one and put it around his neck. When presented with two yellow rosaries, he again grasped the authentic one. Simple objects like canes and quilts were picked out from among the copies. In one instance, given a choice of hand drums, Tenzin ignored the more beautiful damaru (two-headed drum shaped like an hour glass) and chose the plain one, 'Without any hesitation, he picked up the drum. Holding it in his right hand, he played it with a big smile on his face; moving around so that his eyes could look at each of us from close up. Thus, the boy demonstrated his occult powers, which were capable of revealing the most secret phenomena.'[16] The Tibetan bureaucrats were convinced. The Tibetan test relied on supersense powers to detect possession imbued with the essence of the previous Dalai Lama. Tenzin must be the reincarnation as he not only recognized the objects but he could also pick them out from identical copies. But you don't have to be a head of state or religious leader to have the ability to empower possessions with essence.

MISTER ROGERS' CARDIGAN

The longest-running American public television show was *Mister Rogers' Neighborhood*, which began airing in 1968 and had its final episode in 2001. It always began the same way: the congenial Fred Rogers returning home and singing his theme song, 'Won't You Be My Neighbor?' as he changed into his sneakers and a cardigan. It was a children's show that dealt with the anxieties of growing up and coping with problems and expressing emotion, all delivered in a calm and serene formula that did not deviate over the decades. Fred Rogers, a real-life ordained Presbyterian minister, was a homely, placid, and comforting figure to the nation's children. With almost one thousand episodes, Mister Rogers became a significant figure for millions of Americans. He received numerous awards and accolades and even had an asteroid, '26858 Misterrogers', named after him. On receiving a lifetime achievement award at the 1997 Emmys, Mister Rogers brought the audience to tears with his simple and humble acceptance speech. On his death, the US House of Representatives unanimously passed Resolution 111 honouring Mister Rogers for 'his legendary service to the improvement of the lives of children, and his dedication to spreading kindness through example'. When his old car had been stolen and he filed a police report, there was public outrage. Apparently, the car was returned to the same spot with a note saying, 'If we'd known it was yours, we'd never have taken it!' Whether this tale is true or not does not really matter. People would like to believe that it was true. The man was loved by generations.

The iconic symbol of Mister Rogers was his trademark cardigan. Over the course of his career, he actually wore twenty-four, knitted by his mother. One of those cardigans is now on display at the Smithsonian Institution's Museum of American History. Such is the reverence for Mister Rogers. No one could be more different from Fred West.

In a study of supernatural contagion, researchers wanted to know whether children would consider the cardigan of Mister Rogers a special garment imbued with his goodness.[17] First they were shown two identical cardigans and told that one had belonged to Mister Rogers. They were then shown a picture of another child who did not know that one cardigan belonged to the famous man, and they were asked if wearing each cardigan would make the child look, feel, or behave differently. The youngest children, the four- to five-year-olds, did not think that wearing Mister Rogers' cardigan would have any effect. Yet when asked the same questions, six- to eight-year-old children showed the beginnings of magical contagion by saying that the child would behave differently and feel more special. They also thought that something of Mister Rogers would pass from the cardigan.

The most remarkable result, however, was not from children but rather from twenty, mostly female, adult students. Most thought there would be an effect of wearing Mister Rogers' cardigan in a child who did not know who it belonged to. Four out of five adults thought that the essence of Mister Rogers was in the garment, even though they themselves did not particularly want to wear it. This shows that there is a developing supersense when it comes to positive contamination that is a reverse of the Fred West cardigan effect. Both good and evil are perceived to be tangible essences that can be transmitted through items of clothing and contaminate them, and this belief strengthens as we grow older.

THE PRESTIGE

One of my favorite recent movies is *The Prestige*.[18] It's the story of two rival Victorian magicians who try to outperform each other with the ultimate stage illusion known as 'the Transported Man'. Both perfect variations where the magician is apparently instantly transported from one location to another. 'The prestige' is the illusory effect. The two men achieve this effect in different ways. One of them, Alfred Borden, uses the same principle of our copy box experiments and has his otherwise unknown identical brother appear at the second location just at the right time so that it looks like the first brother has been instantly transported. The other magician, Rupert Angier, uses his wealth to recruit the brilliance of the mysterious and enigmatic Nikola Tesla, a real-life maverick genius of the time, to build him an actual copying machine that duplicates the magician at the second location.[19] In the movie, Tesla achieves what our copy box only pretends to do.

Of course, it's a work of fiction, but theoretical physicists have argued that it could be possible to teleport an object by decoding its physical information at one location and sending that information to reconfigure matter at the other end. This would create two versions of the object. Duplication is all very well and fine for inanimate objects, but what about copying real people? How would we cope with an identical copy of ourselves? In the movie, Rupert Angier solves the problem by drowning the original each time he duplicates himself. I think that this is an unlikely scenario. Not many people would willingly kill themselves so that an identical copy could live on for the rest of their lives. Still, *The Prestige* raises really interesting questions about duplicated bodies and minds.

We cannot easily conceive of ourselves as being copied exactly. This stems from our increasing sense of dualism that I described in chapter 5. Physical states may be copied, but not mental ones. As children, we understand our own minds before we appreciate that others also have unique minds. With development, we become aware of our own minds as being unique and what makes us who we are. The possibility of exact duplication of our own minds is an affront to the sense of self. If we consider ourselves unique – and let's face it, we all do – then the possibility that someone else shares exactly the same mind would mean that we no longer have our own unique identity. We would be exact clones. This is why having an identical twin who looks the same is a bit weird but ultimately not a problem. However, having a twin with exactly the same mind would be. We are happy to consider simple animals such as aphids as clones because we generally don't attribute minds to insects. The difference arises with animals that we think may have minds. This is much more worrying, and the reason human cloning is so repulsive to most.

We decided to investigate the origins of these intuitions with a live animal that we apparently instantly copied.[20] We introduced six-year-olds to our pet hamster. We told the children three invisible physical things and three mental states about the hamster. We said that the hamster had a marble in his tummy, had a blue heart, and had a missing tooth. Note that we chose three properties that could not be directly seen. We did this because we wanted to compare invisible physical properties with three mental states, which are, by their very nature, invisible to others. We then induced three mental states of mind in the hamster. We asked the children to tickle the hamster, show it a picture they had drawn, and whisper their name in its ear. Each child understood that the hamster would remember each of these events. We then placed the hamster in the copy machine. When the second box activated, it was opened to reveal a second identical hamster. The question of interest was: which, if any, of the invisible states would the child think were present in the second animal?

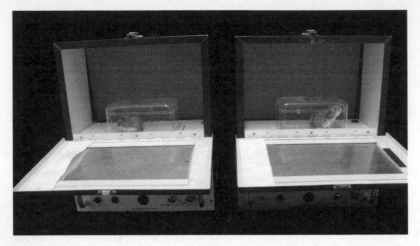

FIG. 21: Then there were two. The copy box machine apparently duplicates a living hamster. AUTHOR'S COLLECTION.

One-third of the children thought the second hamster was absolutely identical on all properties, and one-third thought it was completely different, sharing none of the same properties. The remaining children reported that while the invisible physical properties had copied (the missing tooth, the blue heart, and the marble in the tummy), the mental states had not. Children were beginning to draw a distinction between physical and mental properties and the possibility of duplication. Just to check, we repeated the study with a digital camera that recorded events such as hearing a name and seeing a picture. We also said that the camera contained blue batteries, had a marble inside, and had a broken catch. When we produced a second identical camera from the copying machine, all of the children thought that all properties had been readily duplicated. Likewise, if the original hamster was simply 'transported' from one box to the other, children thought that everything remained intact. Duplication was the problem.

What do these findings tell us? First, children believe that the machine can copy objects faithfully but are less inclined to believe that this is true in the case of a living hamster. They draw a distinction between duplicates of inanimate objects and living animals. In particular, most children think that a copied animal would be different from the original. If anything is copied, it is more likely to be something physical rather than something mental. This suggests that children see living things as more unique than artefacts on the basis of nonphysical properties. This fits with the quiggle study I described earlier.

What if we had copied a real person? I bet that most of the children would not have regarded the copy as having the same mind. After all, would you? Paul and I are still considering whether we would ever be allowed to duplicate a child's mother. Whether this study is ever conducted or not, we strongly predict that children would not readily accept a duplicated mother as a suitable replacement for the original any more than they would accept a copied attachment object. That's because people are also seen as having a unique essential identity. Let me end with a warning about what happens when we lose our capacity to essentialize the world.

CAPGRAS SYNDROME AND THE ALIEN REPLICANTS

When I was a kid, I used to dismantle my toys to see how they worked. It's something that many inquisitive kids do. The way something breaks down can give clues to how it works in the first place. Neuropsychologists are intrigued by how the mind works. They don't go about dismantling minds, but they are very interested in broken minds. The way the mind disintegrates following damage or disease of the brain can be a really insightful way to gain an understanding of normal functioning. We know that damage to certain parts of the brain produces characteristic changes in the mind. It's one of the reasons most psychologists are not dualists: they are very familiar with the idea that the mind is a product of the brain.

One of the more bizarre disorders that is relevant to thinking about the true identity of others is Capgras syndrome.[21] This disorder is a delusional state in which the sufferer typically believes that family members have been abducted and replaced with identical replicants.

Thankfully, the disorder is very rare; only a handful of cases have been reported in the literature. The delusion is associated with paranoia and can be very dangerous. Sufferers have been known to kill 'imposters'. In one extreme case, a sufferer who thought his father had been replaced by a robot decapitated him, looking for the batteries and microfilm inside the head.[22]

Although the delusions usually involve significant family members, reportedly they have applied to family pets and personal inanimate objects as well. One patient thought that his poodle had been replaced with an identical dog.[23] Another woman thought her clothes were replaced by items belonging to other people and would not wear them because she feared the objects would transmit an illness to her.[24] When Capgras patients look in the mirror, they often don't recognize themselves. One husband had to cover every reflective surface in the house because his wife suffered from Capgras syndrome and thought there was another woman out to replace her and steal her husband.[25]

I think that Capgras syndrome is what goes wrong when we lose our supersense that there are essences inside people, pets, and objects.[26] It is more commonly associated with significant others because these are the individuals with whom we are most emotionally connected. One theory for the syndrome explains that our recognition systems for things work by linking the way something looks with an emotional tag.[27] So you get a warm feeling when you look at your spouse, your pet dog, and maybe even your favourite car. When we look at significant others, we not only visibly recognize them but feel them as well. Like normal people, Capgras sufferers remember how they used to feel about such people and items, and they expect to get that same emotional signal.

The problem in Capgras syndrome is that this emotional tag is missing from the process, and so the sufferer is left with only the visible information. So the Capgras sufferer cannot feel that these are the same people, pets, and things that he or she used to experience before the illness. The only logical answer must be that these are not the same people, pets, or things. Rather, they must be identical copies. It's the only way for the Capgras patient to make sense of the experience.

This leads to the paranoid delusion that there is a conspiracy to replace things in the world.

Capgras syndrome is one specific illness out of a range of disorders in which patients believe things are not what they seem. These dissociated disorders reveal how important it is to have an essential perspective on the world. Without this essential sense of identity, people think that the world is a charade. It may look normal, but it lacks emotional depth. Those suffering from Fregoli's syndrome, for instance, believe that someone else has taken on a different appearance. In the even more dissociated disorder known as Cotard's syndrome, patients believe they must be dead because things are not what they used to feel. The world no longer seems real. Ironically, one reason why these syndromes are so rare is that brain injury to those areas that produce these disorders are usually fatal. Those who survive can have their experience of reality fundamentally distorted. The 'something there' that William James talked about has gone. The supersense is part of this connectedness that we all experience, even though we are not fully aware of how it shapes the way we view the world. Without it, experience loses a vital dimension.

WHAT NEXT?

How can we best explain the emerging picture I have sketched here? As we discussed earlier, young children are essentialist in their reasoning about living things. They infer hidden energies and properties to living things from early on, even though they are not taught to think this way.

However, inanimate objects can also take on essential unique properties. In particular, the first sentimental objects may be the ones that help us through the early stages of separation and being left alone as infants. These attachment objects probably soothe children by offering some familiarity each time they are placed alone to sleep. However, over the next couple of years the child becomes emotionally attached to the object. What may have started off as a simple object soon becomes irreplaceable. In the case of attachment objects, it's as if there is an additional invisible property that makes it unique.

Maybe this is where our sense of authenticity comes from, because around the same time children begin to appreciate that certain objects said to belong to significant others have an intrinsic value over and above their material worth. In our study it was the Queen's cutlery and cups, but it could have been Dad's watch or Mum's clothing. I think that this makes sense from the psychological essentialism perspective. In the same way that children's notions of contagion develop, their essentialist beliefs may also change from a localized focus of identity to one that spreads. Somewhere around six or seven years of age, children start to think that certain objects that were previously owned by significant others take on the properties of that person. This not only explains the origin of memorabilia collecting but also the emerging fear of coming into physical contact with killers' cardigans or other conduits of evil. What's more, this attitude may even intensify as we grow into adults and apply essential reasoning to others in the world.

The increasing tendency toward psychological essentialism may be a result of children developing a better understanding of what it is to be unique and an individual. Arguably, as we develop into adults, we have much more sophisticated ways of thinking about others as we form many more categories with which to pigeonhole people. Also, as we saw in chapter 5, children have an increasing sense of the importance of the mind as a unique property of the individual. This is why duplication of minds with a copying machine is so unacceptable. Our natural way of thinking about ourselves and other people leads to an increasing reliance on beliefs about identity, uniqueness, and things that can and cannot connect us.

So I would argue that the behaviour of the toddler towards his grubby blanket and the obsession of a fanatical collector with owning original memorabilia reflect the same human tendency to see objects as possessing invisible properties that originate from significant individuals. By owning objects and touching them, we can connect with others, and that gives us the sense of a distributed existence over time and with others. The net effect is that we become increasingly linked together by a sense of deeper hidden structures.

You may disagree with this theory. You might argue that not all of us form emotional attachments to objects or even collect. How could such a theory apply to the whole of humankind? I would reply that, like many aspects of human personality, such behaviours and beliefs probably exist on a continuum. Some of us are more inclined to this way of thinking than others, but we can all appreciate that there are hidden properties to the world. Like the supersense, we all vary in how far we are prepared to believe that there are additional dimensions to reality. And maybe these individual differences have something to do with the way our brains are wired as much as the different cultures we grow up in. Our supersense may have a biological basis, which I explain in the next chapter.

CHAPTER NINE

THE BIOLOGY OF BELIEF

SUPERNATURAL BELIEFS ARE not simply transmitted by what people tell us to think. Rather, I would argue that our brains have a mind design that leads us naturally to infer structures and patterns in the world and to make sense of it by generating intuitive theories. These intuitive theories create a supersense. I think this happens early in development even before culture can have its major influence. That effect of culture may occur much later in a child's development. Meanwhile, there is something in our biology that leads us to belief. Yes, we can believe what others tell us, but we tend to believe what we think could be true in the first place. How can we prove such an account? The answer is to find a supernatural belief that most people hold but one that does not have its origins in culture. To do that we have to look behind us.

Have you ever felt the hairs standing up on the back of your neck, had the feeling that you were being watched, and turned around to find that someone was indeed staring at you? I don't think there is a single person on this planet who has not had this experience. It's so common that to not have had this experience would in itself be very strange. This sense of unseen gaze has kindled romances and saved lives. Lovers' eyes have met across crowded rooms, and soldiers have turned around just in time to avoid the enemy sniper behind them.[1] It is clearly an ability that has great adaptive value. If only it were true.

People report that they can detect someone looking at them even though there is no way that our natural senses could register this. We can't see them, hear them, smell them, taste them, or feel the touch of their gaze, but people just seem to know when they are being watched. Around nine out of every ten people have this ability. Or at least they believe they do. Stop for one moment and consider how amazing such an ability would be if it were really true.

The sense of being stared at is an example of a common supersense that we have all experienced. In fact, it is so common that it leads to the belief that detecting unseen gaze is a normal human ability. Many educated adults who should know better do not even recognize that such a belief would be supernatural if true. This is why sensing unseen gaze is worth examining in detail as an example of a belief that emerges spontaneously over the course of development but then becomes accepted common wisdom. We don't teach our children this common belief.

If we do not teach the belief of unseen gaze to children, where does it come from? To answer this, it is worth considering some related questions. How does vision actually work? How do we see objects in the world? Is there some energy that leaves the eyes when we gaze at something? The Greek philosopher Plato and the mathematician Euclid believed that vision involves such an 'extramission' of energy from the eyes, a bit like Superman's super-vision.[2] Rays exit the eyes like a torch beam illuminating a darkened cave. Plato even talked about an essence exiting the eyes. However, we have known since at least the tenth century that vision works by light entering the eyes from the outside world as an 'intromission', not the other way around.[3] Light can be reflected from our eyes, which explains the irritating 'red eye' you get from flash photography and the spooky look of cats' eyes caught in the car headlights.[4] However, no modern vision scientist believes that there is energy originating and emanating from the eyes.

That's why you can't see anything when the room lights are turned off or the flashlight is broken. Somehow, such commonsense knowledge doesn't seem to have affected our beliefs. We may understand that sunglasses protect our eyes from harmful light, and yet we still intuitively think of vision working the other way around. Most people, including university students who have taken lessons in optics, believe that vision is the transfer of something entering the eyes at the same time as something exiting the eyes.[5] This probably explains why the sense of being stared at seems so intuitively plausible. If there is something leaving the eyes, then maybe we can detect it. However, there is no current scientific framework that could explain such an ability. It is truly a supersense.

FASCINATING FORCES

'Fascination' means the enchanting power of another's gaze that we find captivating. The psychoanalyst Sigmund Freud used the term in 1921 to describe the power of love, but he drew heavily on ideas from classical mythology and supernatural beliefs.[6] For example, the Medusa was a female monster capable of turning men to stone by a look, and to this day many cultures still have beliefs in the malevolent power of 'the evil eye.'[7] This is the curse that someone can place on you simply by a look. Whenever he was addressing crowds and thought that someone might be giving him the evil eye, the Italian fascist dictator Benito Mussolini reputedly used to touch his testicles as a way of protecting himself. If you find this act a little embarrassing, or don't possess a pair of testicles to touch, magical amulets are still available to protect against the evil eye in Mediterranean countries such as Turkey and Greece.[8]

The Italian Renaissance writers, such as Petrarch (1304–74) and Castiglione (1478–1529), described the look of love (*innamoramento*) as the transfer of particles from the lover's eyes into the eyes of the beloved, which then work their way towards the heart.[9] Here we have the combination of a naive theory of vision working with essences to explain fascination. Our language is peppered with such examples and metaphors that reveal how we treat gaze as something physical that exits the eyes. We talk about piercing eyes or exchanging glances as if there were some physical thing that passes between people.

In the early days, some scientists believed that extramission was a measurable energy force that could be studied in the laboratory. In a paper published in *The Lancet* in 1921, Charles Russ wrote:

> The fact that the direct gaze or vision of one person soon becomes intolerable to another person suggested to me that there might be a ray or radiation issuing from the human eye. If there is such a ray it may produce an uncomfortable effect on the other person's retina or by collision with the other person's ray; it is a fact that after a few seconds the vision of one or the other will have to be turned away at least for a short time. Numerous everyday observations and experiences seem to support the possibility of the existence of a ray or force emitted by the human eye, and in order to give my theory the support of some experimental evidence I decided to try and find or create some instrument which should be set in motion by nothing more than the impact of human vision.[10]

There are many things I find visually intolerable about other people that make me feel uncomfortable and in need of turning away, such

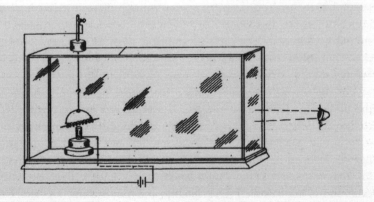

FIG. 22: A reproduction of the patent for a machine to measure the energy of gaze emanating from the human eye filed by Dr Charles Russ in 1919. AUTHOR'S COLLECTION.

as seeing someone pick his nose or clear his sinuses, but I would not make the mistake of assuming that just because another person affects me in a physical way, there is a physical energy field at work. Such logic did not dissuade Russ, who patented a box that contained a copper wire set across a magnetic field to measure this fascinating force.

I couldn't find evidence of Russ's findings being pursued, so we must conclude that the scientific community abandoned his line of inquiry. Bizarrely, this lack of scientific credibility did not deter members of the US military Special Forces from conducting studies to see if goats (and eventually enemy soldiers) could be killed by simply staring at them.[11]

THE SENSE OF BEING STARED AT

In 1898 Edward Titchener reported in the prestigious journal *Science* that nine out of ten of his Harvard undergraduate psychology students believed they had the sense of being stared at.[12] I repeated this survey with more than two hundred Bristol students one hundred years later.[13] To my surprise, the same number of students agreed that it is possible for people to detect unseen gaze, even though these students had taken courses in vision and knew that vision is an intromission process. They should have known that such an ability is scientifically implausible, yet their intuitions told them otherwise. Nevertheless, just because we believe we can sense being stared at does not make it real.

For the record, there are studies that report significant evidence for the ability to detect unseen gaze. A typical way of measuring this is to have an observer stand behind a blind-folded participant, then either stare at the participant or keep his or her eyes closed. Some studies have even been conducted via a camera link, with the two individuals in separate rooms. (This would make Russ's energy field explanation even more implausible.) The staring and not staring are alternated. Trials are repeated many times, and the number of correct guesses is compared to the statistical average of 50 per cent that would be expected if we had no ability to detect when someone is staring at us. The largest study involved eighteen thousand trials with children, and it reported a highly significant effect.[14] Something is definitely being detected here. Isn't that proof enough of the ability?

In my opinion, one of the most interesting discoveries to emerge from these studies is not the ability to detect unseen gaze but rather the remarkable capacity of the brain to detect patterns. Studies that report a significant sense of being stared at have tended to use sequences that may not be truly random. What appears to be happening is that the blindfolded participant is learning to detect these nonrandom sequences.[15] Remember the example in chapter 1 of pressing '1s' and '0s' on the keyboard? Humans are tuned to detect patterns of alternation, even when we are not consciously aware that we are doing so. We seem to be able to detect patterns of sequences if we are given feedback on every trial. If you don't tell participants how they are getting on after each trial, the effect disappears again and performance returns to chance.[16]

Science cannot categorically prove that the sense of being stared at is not true or will never be true in the future, but the evidence is so weak or nonexistent that it must be regarded as unproven. There have been many failures to replicate the effect reliably, so as the saying goes, 'One swallow doesn't make a summer.' It is unscientific to keep flogging a dead horse if the effect you seek refuses to replicate reliably. Not only must scientists find evidence for their theories, but they must also abandon them when the evidence fails to stand up to scrutiny, especially if those theories would overthrow the conventional theories that up to that point have been so reliable. Why should a glimmer of some possible effect overturn a body of work that has undergone rigorous testing and validation? As the saying goes, 'Extraordinary claims require extraordinary evidence.'[17] So where does this common belief in the ability to detect unseen gaze come from?

DEVELOPING A SENSE OF THE UNSEEN GAZE

I think the sense of being stared at is a supernatural belief, but one with a very natural origin that can be traced to a naive theory of how vision works. The sense develops into a full-fledged supernatural belief as we become more tuned in to the language of the eyes and our growing sense of connectedness as adults.

If you ask young children how seeing works, they respond that something leaves the eyes.[18] For example, if you show them a picture of a balloon and a person and then ask children to draw 'seeing', they typically produce an arrow from the eyes to the balloon. Is this so surprising? After all, we look *at* things in the world. We are the source of the looking, so seeing must work that way around. We look by moving our gaze around the world to see the different sights. We control where we look, and so the experience of vision is that it originates from within ourselves.[19] Remember the Numskulls inside our heads guiding our body and controlling our eyes by moving them around to see? It is easy to understand why most of us think seeing works in this way from an early age.

Do such naive beliefs explain the sense of being stared at? Actually, the picture is much more interesting. If you ask children about whether they can sense being stared at, they generally report much lower levels than adults.[20] I expect that's because most young children are so self-centered that they are mostly oblivious to others around them. That's something that changes as we become more self-conscious about being watched. So the sense of detecting unseen gaze actually increases as we get older! Why do more adults than children believe that they can detect unseen gaze? After all, adults should be more scientific and rational than children. I think the explanation involves our increasing social connectedness to others, our attention to their eyes, the developing mind–body dualism we discussed earlier, and the accumulation of evidence that confirms our intuitive beliefs.

THE EYES HAVE IT

Studies of child development reveal that we become much more sensitive to other people's gaze as we get older.[21] Gaze is such an important channel of communication that we automatically pay attention to it.

In fact, we can't ignore it. That's why having a conversation with someone who repeatedly breaks fixation or glances off is so annoying: they are thwarting our attempts to read their thoughts based on their gaze. So gaze is crucially important to us.[22] When someone stares at us, it directly stimulates the emotional centres deep inside our brain. Staring is not a passive act but an active event that affects us emotionally.

The amygdala and ventral striatum are the emotional structures deep within the brain that fire during social exchanges.[23] They give us the feelings we experience during social interactions. Direct gaze at a distance is fine for recognizing people, but direct gaze close up can make us very uncomfortable.[24] If it's coming from a lover, it can make your heart pound and release butterflies in your stomach. If it's coming from a stranger, your mind races (*What does he want with me?*). That's why no one stares at other people inside elevators. We prefer to look at the ceiling or floor rather than each other. We are too close for comfort.

Children, on the other hand, have to be told not to stare. As we saw earlier, babies look at eyes from the very beginning, but with age, we become more attuned to gaze. As we approach adulthood, we need to be able to figure out friend or foe, and so we increasingly learn the subtleties of social interaction and the meaning of a glance. We also become more self-conscious about the others around us, and our need for social approval intensifies. Anyone who has ever been to a party of adolescents can't fail to notice the flurry of exchanged glances between the sexes. These fledgling adults are embarking on the first stages of intimacy, and these early steps involve reading the language of the eyes.[25]

The emotional arousal we experience when we are being stared at simply reinforces our intuitive sense that we can detect another's gaze as a transfer of energy (*Why else would I feel this way when she stares at me?*). Now put yourself in a situation where you suddenly feel uncomfortable with other people around. With this naive theory, we readily remember every occurrence when we sensed this discomfort that proved to be justified, but we conveniently forget every time when we were wrong. Like any theory, this one comes with a bias to seek out confirming evidence of what we think is true in the first place.

This tendency to look for confirming evidence is known as the confirmation bias. It's the prejudiced reasoning we exercise whenever we make judgements that fit with our preconceptions. We rarely take things at face value but rather look for confirmation of what we believe to be true in the first place. This has been used to great comic effect in the US by the American mortgage company Ameriquest, which has been running an ad campaign showing how easy it is to jump to unjustified conclusions when you don't know all the facts and when you reason according to your preconceptions. My favourite is the father who is giving his daughter and her friends a lift when they stop off to buy some chewing gum. He calls her back briefly to the car to give her money to buy the gum. As she leans in through the window, he says, 'Here's some money.' At that moment, a police patrol car pulls up behind. 'What have we got here?' says the officer as the older man is handing over money to the clearly underage girl. 'I'm her daddy,' stutters the father caught in the headlights like a startled deer. The tagline is: 'Don't judge too quickly. We won't.'

The confirmation bias reveals that preconceptions easily shape the way we interpret information. If you think that you can detect unseen gaze, then you remember every example that confirms your belief and conveniently forget all the times you were wrong.

Finally, a sense of being stared at can strengthen from the error of causal reasoning, *post hoc, ergo propter hoc* (after this, therefore because of this), described earlier as the basis for superstitious reasoning – in other words, assuming a cause where there is none. Imagine the situation. You are walking down the road and pass a group of youths. You get the uncomfortable feeling that they are staring at you. You stop, turn around and find out that you were right. But consider the sequence again from the perspective of one of the youths. You are hanging out with your friends and this guy walks past. You give him a glance but continue talking to your friends. Suddenly the guy stops and turns around. What do you do? You look back at him to see why he has turned around. In other words, we may think that we turn around because we sense others looking at us from behind but, in reality, they look at us because we turned around to face them in the first place. We are so self-conscious and socially sensitive that this sort of event must happen all the time. Such episodes simply reinforce, however, our beliefs that we can detect when we are being watched.[26]

Of course, I may be wrong, and billions of people will disagree with me. After all, they have all had personal experience of the phenomenon and that's why people believe in the supernatural. But, like the invisible square we saw in chapter 1, just because we all experience something does not make it real. The most prominent and active advocate of the sense of being stared at is Rupert Sheldrake, who proposes that this ability reflects a new scientific theory of disembodied minds extending out beyond the physical body to connect together. I regard this as an idea originating from the dualism of mind and body that we discussed earlier, but such a notion has been rejected by conventional science. Undaunted by 'scientific vigilantes', Sheldrake proposes that the sense of being stared at and other aspects of paranormal ability, such as telepathy and knowing about events in the future before they happen, are all evidence for a new field theory that he calls 'morphic resonance'. He proposes that it is similar to other examples of field phenomena in nature such as electric and magnetic fields.[27] His idea is that the scientific evidence for morphic resonance will come from quantum physics, where the natural laws that govern the physical world as we know it no longer apply. This may turn out to be true, but for the moment I do not think morphic resonance qualifies as a field phenomenon.

The trouble is that, whereas electric and magnetic fields are easily measurable and obey laws, morphic resonance remains elusive and has no demonstrable laws.[28] No other area of science would accept such lawless, weak evidence as proof, which is why the majority of the scientific community has generally dismissed this theory and the evidence. However, this has had little influence on the general public's opinion. Science may be wrong about the reality of the sense of being stared at, but what is clear is that the public's belief in the phenomenon is much stronger than the best measures obtained for its existence so far suggest.

BIG BROTHER IS WATCHING YOU!

The sense of being stared at reflects a common concern about being observed and monitored. George Orwell describes a paranoid world in his classic novel *1984*, in which every action and belief of the citizens is controlled by the thought police overseen by the eyes of Big Brother.[29] We tend not to engage in crime when we are being watched. For obvious reasons, we prefer to remain undetected. That's part of the thrill of shoplifting by those individuals who steal items they can readily afford. The excitement is the reward, not the actual object. If we are being watched, we generally conform to social rules. People even become overtly social and more cooperative when they know they are being observed.[30]

Have you ever felt that pang of guilt when you have done something wrong and then wondered whether someone saw you doing it? It doesn't have to be a real person watching you. For example, honesty boxes depend on the virtue of people to own up and pay for something if they have used it. Typically, these are the boxes in staff rooms and clubhouses that rely on members to make a fair contribution towards the cost of something, usually a hot drink. They generally don't work that well unless there is someone watching the partakers. In one study, researchers posted either a set of human eyes or a picture of flowers above the honesty box for coffee and tea.[31] On average, people paid almost three times more into the honesty box during the weeks when a picture of staring eyes was posted compared to the weeks when a picture of flowers was presented, even though there was no difference in how many cups of tea or coffee were poured. The eyes made people feel guilty about not paying for their drinks!

Sometimes the thought of someone watching us from beyond the grave is enough to make us behave ourselves. For example, students found they had the option to cheat on a computer-based exam when, every so often, the computer 'accidentally' gave away the correct answer. In fact, the experimenters had deliberately programmed this to happen because they were really interested in whether participants would cheat by using this information as their answer or behave honestly on the exam. To put the students in the right frame of mind, an assistant casually told them before the test that the exam room was said to be haunted by a former student who had died there. Exam results showed that students who had been told the ghost story were less likely to cheat compared to students given no such story.[32] Our sense of honesty is arguably policed by our feelings of guilt. Part of that guilt comes from the anticipated social disapproval we believe we would experience if we were found to be breaking some rule. Students who believed that a former student might have been present in the exam room were less willing to cheat.

This guilt trip theory has been used to explain why we so readily believe in an afterlife. The psychologist Jesse Bering thinks that the belief in ghosts and spirits may have evolved as a mechanism designed to make us behave ourselves when we think we are being watched.[33] A guilty conscience works because it polices the way we behave, and if it can be easily triggered by the sense of others watching us, then we are more likely to act in a way that is for the benefit of the group. In the same way that students are less likely to cheat when told a ghost story, if we believe the ancestors are watching us, we are more likely to conform to society's rules and regulations. Such a way of thinking, being advantageous to the group, would be likely to be passed on from one generation to the next. As we saw in chapter 5 on mind-reading, assuming the presence of others could be a good evolutionary strategy to always be on the lookout for potential enemies.[34] And if we are hardwired to assume the presence of agents and spirits in the world, even the slightest example of a pattern that could be a face or a pair of eyes will readily be seen as such. Any bump in the night could be another person.

THE MAGIC OF MADNESS

Thinking that others are watching you and talking about you is a classic symptom of psychotic mental illness, most notably paranoid schizophrenia.

Not surprisingly, supernatural beliefs are a major feature of the psychotic disorders of mania and schizophrenia. Mania is characterized by excessive energy and productivity as well as inappropriate social behaviour. Schizophrenia takes a variety of forms but is generally a state in which one holds irrational and paranoid delusions and experiences perceptual distortions of reality, especially auditory hallucinations.

One characteristic of all these psychotic disorders is the sense that there are significant patterns of events in the world that are somehow directly related to the patient. This way of sensing meaningful patterns is known as apophenia, which refers to an abnormal tendency to see connections in the world that are considered relevant by the patient.[35] Apophenia helps to explain the basis of psychotic symptoms such as paranoid delusions of persecution. For example, psychotic patients in the midst of a paranoid episode typically report that there is a conspiracy centered on them. They are certain that they are being watched, that people are talking about them, that their phone lines are tapped, and that generally there is a coordinated hostile campaign against them. For the sufferer, these delusions are very real and beyond rational control.

We can all sense patterns, but psychotic patients are more prone to do so and to interpret patterns as significant events related to them personally. This is supported by research that demonstrates a relationship between sensing patterns and symptoms of psychiatric disorder.[36] Even adults who do not exhibit full-blown psychotic mental breakdowns, the so-called 'borderline' cases, have been shown to hold a strong supersense. These beliefs are called 'magical ideation', and they can be measured by responses to statements such as:

'Some people can make me aware of them just by thinking about me.'

'I think I could learn to read others' minds if I wanted to.'

'Things sometimes seem to be in different places when I get home, even though no one has been there.'

'I have noticed sounds on my records that are not there at other times.'

'I have had the momentary feeling that someone's place has been taken by a look-alike.'

'I have sometimes sensed an evil presence around me, although I could not see it.'

'I sometimes have a feeling of gaining or losing energy when certain people look at me or touch me.'

'At times I perform certain little rituals to ward off negative influences.'

These statements are taken from a 'magical ideation' questionnaire used by researchers to study the relationship between mental illness and the supersense.[37] If you score highly on this questionnaire of thirty items, you are predisposed to psychosis. It does not mean that you definitely are psychotic or will have a psychotic breakdown, but rather that you may be at risk. See how highly you score by completing the questionnaire in the Reader's Notes.

Such aspects of human nature are generally spread out across a population – a bit like height, for example. Some of us are very tall, and some of us are very small, but most of us are in the middle. It's the same with thought processes. Some of us are more intelligent than others. Some are more anxious. Others are more depressed. Magical thinking is just the same. Psychosis can be regarded as one extreme of the distributed range of beliefs. We can all experience episodes of depression, anxiety, delusion, obsession, compulsion, paranoia, and all manner of psychiatric conditions. However, when these episodes start to dominate and control an individual's life, they are said to be pathological. They become an illness that disrupts the individual's well-being.

The items from the 'magical ideation' questionnaire clearly reflect some of the pattern-detecting and intuitive beliefs that I have been describing throughout the book. Normally, we may briefly entertain such notions, but we can readily ignore or dismiss them as irrational. If we have an intrusive thought out of the blue, it does not faze us. We can inhibit the thoughts that form in our mind.

In contrast, psychiatric patients are unable to control these thought processes. They may even attribute such thoughts as coming from some other source. This is why schizophrenics often think their thoughts are being transmitted or invaded by outside signals. Everything is given significance. Consider this example taken from a schizophrenic nurse describing her first psychotic episode. The passage clearly reveals the supersense at work:

> Every single thing 'means' something. This kind of symbolic thinking is exhaustive . . . I have a sense that everything is more vivid and important; the incoming stimuli are almost more than I can bear. There is a connection to everything that happens. No coincidences. I feel tremendously creative.[38]

The supersense is characterized by beliefs and experiences that lead us to infer hidden structures, patterns, energies, and dimensions to reality. We see ourselves as extended beyond our bodies and connected by an invisible oneness of the universe. Without adequate inhibitory control, we would be overwhelmed by our supersense. How do we stop these thoughts?

DOPAMINE: THE BRAIN'S SUPERNATURAL SIGNALLER?

In this book, I have been arguing that the supersense is a natural product of the human brain. However, we all vary in the extent to which we experience the supersense. If it is not culture that can explain these individual differences in the way we interpret the world, there must be something in our biology that can explain this variation. At this point, I apologize to brain scientists around the world for the overly simplistic picture I am about to paint.

The brain works as a collection of cells wired together in networks to process incoming information, interpret that information, and then store it as knowledge. These various tasks are much more complicated than a few sentences can ever describe, but they all depend on networks of connected cells that communicate with each other through minute electrochemical activity. This is achieved by the neurotransmitters that form the signalling system of the brain.

Dopamine is one such chemical neurotransmitter. As the neuroscientist Read Montague says, 'The dopamine system is hijacked by every drug of abuse, destroyed by Parkinson's disease, and perturbed by various forms of mental illness.'[39] Antipsychotic drugs that alleviate the florid delusional symptoms of schizophrenia are known to reduce the activity of the dopamine system, whereas administering dopamine to Parkinson's patients, who already have impaired dopamine production, induces hallucinations and supernatural experiences. For example, in one study the most common hallucination was the sense of someone else in the room.[40] Abuse of illegal drugs such as amphetamines and cocaine can lead to supernatural experiences, and guess what? They affect the dopamine system. For these reasons, dopamine has been a source of interest for those trying to understand the supersense. If there is a smoking gun for the biological basis of the supersense, it seems to be firmly held by the hand of dopamine.[41]

The neuropsychiatrist Peter Brugger has proposed that apophenia represents abnormally excessive activity of the dopamine system that leads individuals to detect more coincidences and see patterns that the rest of us miss.[42] The idea is that the dopamine system acts like a filter. Too much dopamine-related activity in the brain and all sorts of patterns and significance are perceived. Too little and nothing is detected. If you score high on the 'magical ideation' scale described earlier, you are also more likely to detect patterns and sequences than those who score low. In other words, sceptics and believers differ not only in their supersense but also in how they perceive the world. This is an important point that my colleague, Susan Blackmore, has made throughout her life's search for proof of the paranormal. People differ in the way that they interpret the evidence.[43] These differences in the way individuals perceive the world can be illustrated with a visual metaphor. Who do you see in this picture of a famous celebrity over the page?

If your eyesight is reasonable then you will probably recognize Albert Einstein. You have picked up the fine detail of information that depicts the famous scientist. But maybe you are missing the bigger picture? Try screwing up your eyes to blur the image or better still, put the

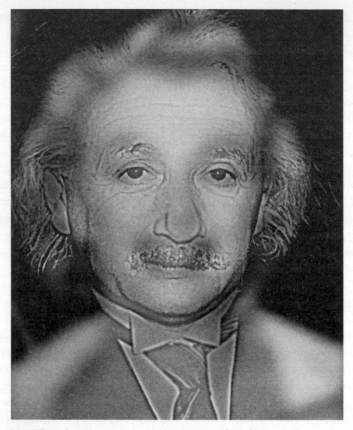

FIG. 23: Who do you see in this picture? © AUDE OLIVA.

book down and walk away about 10 feet. Take another look. Who do you see now? It is exactly the same image but now you see the actress Marilyn Monroe. In the same way that we constantly filter information from our environment for significant patterns, individuals may literally have different ways of viewing the world.

Sceptics and believers may also differ in the activity of their dopamine systems. For example, imagine watching your TV when the antenna is not plugged in. The fuzzy snow on the screen is like visual noise. If you were to put a very faint image of a face against such a background, believers would be much more likely to say that a face was present compared to sceptics, who require more evidence of a face. Sceptics more often reject the presence of a target when it is really there. That's because sceptics and believers have different thresholds. To test this, Brugger and his colleagues asked sceptics and believers to detect words and faces presented on a computer screen among lots of visual noise. The researchers then administered the drug levadopa to raise dopamine levels in both groups. The sceptics now saw patterns, but the believers were more conservative. The dopamine changed the setting on the filter for those in these two groups. Changing levels of the neurotransmitter had altered each participant's perception.[44]

The research into the brain mechanisms of the supersense is intriguing but hardly surprising. We know that reality can be easily distorted by changing brain chemistry. Hallucinogenic drugs induce fantasy states in which all sorts of supernatural beliefs can operate. That's why mind-altering substances and rituals have been so important to religious ceremony. Whether through poisonous plants or trance-induced rapture, altering the brain alters reality.

An altered sense of reality may be the reason why psychotic mania has often been linked to creativity. The tendency to seek and perceive patterns where the rest of us see nothing may be part of the creative process. Some of the world's most creative artists, writers, composers, and scientists have been associated with periods of mania, and many have had full-blown psychotic breakdowns. Listing some of them is like compiling a who's who of the creative world: Van Gogh, Beethoven, Byron, Dickens, Coleridge, Hemingway, Keats, Twain, Woolf, and even Newton – all experienced episodes of mania. Creativity may be a benefit of the supersense, but the price we sometimes pay is potential mental illness.

However, we don't have to suffer from psychiatric illness to assume that the supersense is operating in the world. Rather, sensing patterns and connections is part of the normal process, but we must also learn to ignore patterns and connections that may not really exist. Supernatural thinking may interfere with our ability to act rationally, as when we assume the presence or activity of unseen events in the world when they are not really there. To overcome this problem we need to exercise some form of mind control.

MIND CONTROL

The supersense may result from a mind designed to infer invisible structures in the world, but not all of us succumb to the idea that the supernatural is real. Many of us can ignore such intuitive reasoning. How can this be? Consider again some of the phenomena outlined in this book. Why does a child search over and over again for a fallen object directly below? Why do children have a problem understanding that things that look alive are not really so? Why are children's intuitive theories about how vision works difficult to ignore? Why can we not ignore someone else's gaze? Why might childish intuitive misconceptions lie dormant in the adult only to reappear later in life? Why do we fail to ignore silly thoughts? Why do psychotic patients detect all manner of significant patterns in the world? In all these situations, there is something about how the mind organizes and controls what we do and think. We need mind management to stop ourselves getting stuck in routines and thoughts.

Scientists interested in understanding how the mind works have increasingly become interested in the developing front part of the brain. In terms of sheer size, the frontal parts of the brain are enormously expanded in the human species. This explains why our foreheads are so much bigger in comparison to other primates and prehominid fossil skulls. Unlike our closest animal cousins, we stand out in our ability to plan and coordinate behaviour and thoughts in a flexibly adaptive way. We can anticipate events and imagine solutions. Our frontal brains being what they are, we could easily beat other monkeys and apes at rock, paper, and scissors.

One region of the frontal lobes has been a prime focus of interest: the dorsal lateral prefrontal cortex, or DLPC. The DLPC plays a major role in controlling a set of operations known as the executive functions of the brain, which include:

1. Working memory: The ability to hold temporary thoughts in mind without necessarily committing them to memory.
2. Planning: The ability to anticipate future events and organize a corresponding sequence to achieve goals.
3. Inhibition: The ability to ignore distracting or irrelevant thoughts and actions.
4. Evaluation: The ability to weigh up thoughts and actions in terms of desired goals.[45]

Working memory does exactly what the term implies.[46] It allows you to work out problems by holding on to information in a temporary memory store. You use working memory every time you have to remember a new telephone number or someone's name at a party. Information in working memory is only briefly held in store. It's a store that is fragile and limited. That's why it can be very hard to remember very long telephone numbers unless you rehearse them by repeating them over and over. Working memory is like a temporary back of a mental envelope we use when we want to take note of something briefly.

Planning is how you achieve your goals. It allows you to imagine and build mental models to play out different scenarios in advance. For example, consider this brain-teaser: You have a fox, a chicken, and a bag of corn to transport across a river but you only have enough space in the boat for one item on each trip. How do you transport all three across the river without losing any items? Remember foxes eat chickens and chickens eat corn, so you can never leave any of these pairs alone on the bank. To solve this you need planning.[47] You can imagine the consequence of the first trip, the second trip, and so on until you work out the solution. If you don't know the answer to this one, it requires taking the chicken back and forth over the river more than once.

Inhibition is another important operation of the DLPC. We need inhibition to cancel inappropriate thoughts and actions. For example, quickly say out loud the colour of the ink – black, white, or grey – for the following words as fast as you possibly can.[48]

word word **word** word **word** word

This should be relatively easy. Let's make it easier still.

white **black grey black** white **grey**

Okay, so you're an expert now. Try to say the colour of the word in the next list as fast as you can.

grey black **white** grey **black white**

Did you make any mistakes? Maybe not, but I bet you had a problem and were much slower. The act of reading triggers the impulse to utter the word as read, but if the word conflicts with the correct answer, that response has to be ignored in order to state the colour. On the other hand, naming a colour is not automatically triggered by reading. So saying the word needs to be suppressed or inhibited in order to make the correct response. This is why inhibition is necessary for planning and controlling behaviour: it enables you to avoid thoughts and actions that get in the way of achieving your goals.

Finally in order to benefit from all this executive function, we need to evaluate our performance. As we saw earlier, adaptive behaviour can help us learn from past successes and mistakes. Remember Damasio's frontally damaged patients in chapter 2 who were unable to play the gambling game successfully? They lacked the necessary evaluation of the hidden rules controlling the rewards. The system that learns from the past and helps us to make decisions about the future includes the

DLPC. One of the main neurotransmitter systems of the DLPC is ... yes, that's right ... dopamine. This may all be too convenient and simplistic, and it may be my supersense of connectedness at work, but there does appear to be a coherent pattern emerging.

We now think that brain changes in the DLPC have important implications for child development and advances in reasoning.[49] Control of behaviours, thoughts, reasoning, and decision-making – in short, just about every aspect of higher intelligence that humans possess – is dependent on the executive functions of the DLPC. As we develop into adults, we become increasingly more in control of our urges, and that requires the activity of the DLPC. For example, do you remember falling objects? Which falls faster, a heavy object or a lighter one? We intuitively think that heavier objects should fall faster and are surprised if they don't. When adults learn that this belief is wrong, measurements of their brains while they think about the problem reveal that their DLPC is active.[50] When adults reason about the Linda problem from chapter 3 and consider whether she is more likely to be a bank worker or a feminist, their DLPC is active trying to suppress the tendency to go for the most obvious intuitive answer.[51] Even when they give the correct answer, the old childish naive theories are still active and must be suppressed. Bad ideas don't go away. They hang around and have to be ignored!

However, like many functions of the human body, there is a progressive decline in executive functions towards old age Many of the popular mind puzzles, like Sudoku or the current fad for 'brain training' computer games, tap into DLPC abilities. When they claim that they can measure how old your brain is, they do this by comparing your performance on tasks dependent on the DLPC to the normal range that can be expected for people of different ages. That's because DLPC function changes with age.

One consequence of the loss of DLPC control in an adult is reverting back to behaving and thinking like a young child. Whenever this system is impaired through ageing, damage, or disease, the ability to remember, inhibit, plan, and evaluate is compromised. We forget things. We all know elderly relatives who seem to become socially embarrassing in their lack of control. Planning a trip becomes a chore. We may lose the ability to make rational, balanced judgements and leave all our inheritance money to 'that nice lawyer who has been ever so helpful'. Old age does not guarantee wisdom.

THE CRUELLEST DISEASE

For all too many of us entering old age, there can be a much more devastating and progressive slide into decline as we lose DLPC functions. Alzheimer's disease is often considered the cruellest of diseases. The change in personality is the most distressing aspect of the illness. Someone you have spent your life knowing and loving turns into a complete stranger who needs the attention and care of a small child. Alzheimer's is a neurodegenerative disorder, which means that it primarily destroys the higher functions that control behaviour and thinking. It starts off with absentmindedness. Then there are unprovoked violent outbursts, and inappropriate behaviour can alert family members that things are not quite right. The problem with diagnosing the onset of Alzheimer's is that as we age we all change in our personality. We can become forgetful, disinhibited, grumpy, and so on, but Alzheimer's disassembles the individual to the extent that he or she becomes unrecognizable to family and friends.

Recently, research on Alzheimer's has provided unexpected evidence for the supersense. Before adults with Alzheimer's reach a state of advanced decline, they display signs that the mind never truly abandons childish ways of reasoning.[52] For example, when asked, 'Why are there trees?' 'Why is the sun bright?' or 'Why is there rain?' patients give answers just like young children. They say trees are for shade, the sun is bright so that we can see, and rain is for drinking and growing. They have gone back to the teleological thinking of the seven-year-old we saw in chapter 5. They also become animists again, attributing life to nonliving things like the sun. It's not the case that they have forgotten everything they know.[53] Rather, the errors they make reflect the intuitive theories of children. Dementia shows that intuitive thinking is not abandoned but suppressed by the higher centres of the brain as we grow into adults. When that ability to inhibit is lost, the intuitive theories reappear.

BEING IN TWO MINDS

Psychologists have come to the conclusion that there are at least two different systems operating when it comes to thinking and reasoning.[54] One system is believed to be evolutionarily more ancient in terms of human development; it has been called intuitive, natural, automatic, heuristic, and implicit. It's the system that we think is operating in young children before they reach school age. The second system is one that is believed to be more recent in human evolution; it permits logical reasoning but is limited by executive functions. It requires working memory, planning, inhibition, and evaluation. This second reasoning system has been called conceptual–logical, analytical–rational, deliberative–effortful–intentional–systematic, and explicit. It emerges much later in development and underpins the capacity of the child to perform logical, rational problem-solving. When we reason about the world using these two systems, they may sometimes work in competition with each other. The rational system is slow and ponderous. It's not very good at coming up with snappy decisions. Also if you preoccupy your rational system with problem-solving that uses up your executive functions, then the intuitive mechanisms can run amok. That's why people under stress and time constraints often default to the intuitive system that is more effortless. When this happens we make all sorts of supersense judgements.

The supersense we experience as adults is the remnant of the child's intuitive reasoning system that incorrectly comes up with explanations that do not fit rational models of the world. One might assume that those prone to the supersense and belief in the paranormal are lacking in rational thought processes, but that would be too simplistic. Studies reveal that the two systems of thinking, the intuitive and the rational, coexist in the same individual. There are, in effect, two different ways of interpreting the world. In fact, when we measure reliance on intuition, no relationship has been found with intelligence. Intuitive people are not more stupid.[55] They are, however, more prone to supernatural belief. One recent study found that mood is an important factor in triggering supernatural beliefs in those who score more highly on measures of intuition.[56] For example, happy, intuitive adults are more likely to sit farther away from someone they believe is contaminated, a response that reflects the psychological contamination we described in chapter 7. They are also less able to throw darts at pictures of babies; this measure reflects the sympathetic magical law of similarity by which objects that resemble each other are believed to share a magical connection. Even though individuals may not be consciously aware of the thought processes guiding such behaviour, these effects reveal a deep-seated notion of sympathetic magical reasoning. The supersense lingers in the back of our minds, influencing our behaviours and thoughts, and our mood may play a triggering role. This explains why perfectly rational, highly educated individuals can still hold supernatural beliefs.

Marjaana Lindeman at the University of Helsinki has recently tested this dual model of belief and reason and the role of naive intuitive theories.[57] She investigated intuitive reasoning and the supersense in more than three thousand Finnish adults. First, she asked them about their supernatural beliefs, both secular and religious. Then she assessed their intuitive misconceptions. She asked them questions about animism, teleological reasoning, anthropomorphism, vitalism, and core conceptual confusions they had about physical, biological, and psychological aspects of the world − all the sorts of areas that children naturally reason about by themselves that sometimes lead to misconceptions. She asked questions like. 'When summer is warm, do flowers want to bloom?' or 'Does old furniture know something about the past?' Finally, she asked them which style of thinking they preferred − intuitive gut reactions or well-thought-out analytical reasoning.

When she compared adults with a strong supersense with those who were more sceptical, Lindeman found that believers were more likely to misattribute properties of one conceptual category to another.

For example, they were more likely to say that old chairs know something about the past (attributing mental property to inanimate objects) or that thoughts could be transferred to others (attributing physical properties to mental states). They were teleologically more promiscuous and inclined to animism as well as anthropomorphism. They were also more vitalist and had a sense that things are connected in the world. Were they less educated? No. These were university students. What's more, they scored just as high as the sceptical students on other measures of rationality. Rationality and supernatural beliefs can coexist in the same individual. These students were SuperBrights who simply preferred, or were more inclined to rely on, their intuitive ways of thinking.

Finland may have one of the highest rates of atheism in the world, but this large study of adult students proves that educated people do not neatly divide into those with a supersense and those without one. When people rely on their fast, unlearned gut responses, they are inclined to use their supersense, and it's something that is easily triggered in most of us.

WHAT NEXT?

When I was a child, I spoke as a child, I understood as a child, I thought as a child: but when I became a man, I put away childish things.

— Corinthians 13:11

Throughout this book, I have been arguing that beliefs in the supernatural are a consequence of reasoning processes about natural properties and events in our world. This includes a mind design for detecting patterns and inferring structures where there may be none. Our naive theories form the basis of our supernatural beliefs, and culture and experience simply work to reinforce what we intuitively hold to be correct. This is why the sense of being stared at is such an interesting model for the origin and development of supernaturalism. Children are not told that humans can detect unseen gaze. In fact, it's not something they readily report that they can do. Nevertheless, young children and many adults think that vision works by something leaving the eyes. So when they experience episodes of seeming to detect unseen gaze, this belief simply emerges naturally as an unquestioned ability. It is not even considered supernatural by most people. Children were not told to think this. This model shows how the combination of intuitive theories, pattern detecting, and eventual support from culture produces a universal supernatural belief.

I think that something very similar may be going on for other supernatural beliefs. The notion of psychological contamination we examined in earlier chapters emerges naturally out of psychological essentialism, which has its roots in our naive biological reasoning. Again, this way of thinking is not something that we teach our children. Intuitive dualism and the idea that the mind can exist independently of the body is another. All of these ways of thinking are both naturally emerging and yet supernatural in their explanations of the world.

As we noted earlier, some have argued that adult supernaturalism is a product of religious indoctrination of our children. However, I hope I have convinced you that the various secular supernatural beliefs we have examined throughout this book seem to arise spontaneously without necessarily being started by religion. Most importantly, some beliefs remain dormant, whereas others that are not regarded as supernatural grow in strength. This occurs even in highly educated adults. We can all entertain weird and wonderful beliefs about the world.

We may put away childish things, as Corinthians suggests, but we never entirely get rid of them. Education can give us a new understanding and even progress to a scientific viewpoint, but development, distress, damage, and disease show that we keep many skeletons in our mental closet. If those misconceptions involve our understanding of the properties and limits of the material world, the living world, and the mental world, there is a good chance that they can form the basis of adult supernatural beliefs.

As children discover more about the real world, they should progress to a more scientific view of the world. Clearly, this does not necessarily happen. Most adults hold supernatural beliefs. The supersense continues to influence and operate in our lives. It may even give us a sense of control over our behaviours. As we saw in the opening chapters, many of our actions, whether we are avoiding a cardigan, demolishing a house, touching a blanket, or engaging in exam rituals, give us a psychological way of dealing with things. Without these beliefs, we may feel vulnerable. We may not even be aware that a supersense is influencing our lives, and yet it clearly does.

So, can we ever evolve out of irrationality? Why would such a way of viewing the world continue to flourish in this age of reason? Will the human race ever become ultimately reasonable?

I don't believe so. There is one final piece of the puzzle that I have been hinting at throughout the book that now needs to be considered. It moves beyond the question of origins and asks: are there any benefits of the supersense? After all, if science has the potential to elevate the human species to new levels of achievement, why do we still succumb to a supersense? Part of the answer is that it may be unavoidable, as I hope you will now appreciate. Another reason is that the supersense makes possible our capacity to experience a deeper level of connection that may be necessary for humans as social animals.

Even though humans have the capacity to reason and make judgements, I think that we will always regard some things in life as not reducible to rational analysis. That is because society needs supernatural thinking as part of a belief system that holds members of a group together by sacred values. In the final pages, I will explain how this supersense forms the intuitive rationale for the sacred values that bind our society together.

CHAPTER TEN

WOULD YOU LET YOUR WIFE SLEEP WITH ROBERT REDFORD?

IN THIS BOOK, I have proposed that humans are compelled to understand the nature of the world around them as part of the way our brains try to make sense of our experiences. This process starts early in childhood, even before culture has begun to tell children what to think. Along the way, children come up with all manner of beliefs about the world, including those that would have to be supernatural if true. These ideas go beyond the natural laws that we currently understand and hence are *super*natural. Whether it is a disembodied mind floating free of the body, a sublime essence that harbours the true identity of people, places, and things, or the idea that people are all connected by tangible energies and hidden patterns, these notions are all intuitive ways of thinking about the world. We persist in these beliefs despite the lack of compelling evidence that the phenomena we think are real do in fact exist. Culture may fuel these beliefs with fantasy and fiction, but they burn brightly in the first place because of our natural inclination to assume 'something there', as William James put it. Culture simply took these beliefs and gave them meaning and content.

If we are deluded, can we ever get rid of such a supersense? Will humankind ever evolve into the Bright species that uses logic over and above emotion and intuition? This seems unlikely for a number of reasons. The first reason, which I have been at pains to labour throughout the book, is that the supersense is part and parcel of our mind design and so is deeply embedded in the way we reason. We may possess the capacity for both logical analysis and intuitive reasoning, but one is slow and ponderous while the other is fast and furious. Intuition is not something we can easily ignore, and although we can learn to think in a rational–analytical way, intuitive reasoning has the advantage in the race to influence our decision-making because it is so effortless, covert, and rapid. When a taxi driver asked the late Carl Sagan, the cosmologist, for his gut reaction to the question of whether UFOs are real, Sagan replied that he tried not to think with his stomach. For the rest of us, such control is often lacking as we succumb to naive intuitive reasoning. It is not always right, but we must remember that it has served us well in the past. Otherwise, as a species, we would not be around to tell the tale. The supersense comes from our intuitive reasoning systems and so is part of our makeup. This brings me to another, more important reason for why we may foster a supersense.

I think the supersense will persist even in a modern era because it makes possible our commitment to the idea that there are sacred values in our world. Something is sacred when members of society regard it as beyond any monetary value. Let me give you an example. Life can be full of difficult decisions. People who run hospitals are constantly faced with choices involving life and death. Imagine that you are a hospital administrator and you have $1 million that can be used to perform a lifesaving liver transplant operation on a child or to reduce the hospital's debt. What would you do? For most people, this would be a nobrainer – of course one must save the child.

The economic psychologist Philip Tetlock has shown that people are appalled when they hear that an administrator would make the decision to benefit the hospital, even though more children would gain in the long term from such astute financial planning.[1] What's more, they are also outraged if the hospital administrator decides to save the child but takes a long time to arrive at that decision. Some things are sacred. You should not have to think about them. You can't put a price on them. Likewise, if the choice has to be made between saving one of two children, this decision must take a long time. The choice should not be made quickly. This unbearable dilemma has become known as

'Sophie's choice', following William Styron's novel about the Jewish mother who was forced to decide which of her two children would die in the Auschwitz gas chambers and which would survive.[2] She chose to let her son live and her daughter die.

We intuitively feel that some things are right and some things are just plain wrong. Some decisions should be instantaneous, while others must be agonized over. Decisions can haunt us even when there really should be no indecision. Every choice has a price tag if we care to consider relative worth. There are no free lunches, and so while we may be outraged and indignant about some choices and decisions, the reality is that all things can be reduced to a cost–benefit analysis.

However, a cost–benefit analysis is material, analytic, scientific, and rational. This is not how humans behave, and when we hear that people think and reason like this, we are indignant. When Robert Redford's character offered $1 million to sleep with Woody Harrelson's wife (Demi Moore) the movie audience knew that it was an *Indecent Proposal*. It was morally repugnant. Better that she should have had an affair than do it for money. If you love someone, no amount of money should enter the negotiation, even if he does look like Robert Redford! For many, this $1 million decision is much easier than the hospital administrator dilemma. Likewise, when we hear that people could wear a killer's cardigan, live in a house of murder, or collect Nazi memorabilia, we are disgusted. We feel it physically. Though a cost–benefit analysis may reveal our reaction to be out of balance with the actual costs, we still intuitively feel a moral outrage and violation of society's values.

This is because humans are a sacred species. We treat sacred places, sacred objects, and sacred lives as beyond commercial value. The value placed on each depends on who is making the decision, but each sacred thing could literally be 'priceless'. The alternative is to accept that everything has a price.

The trouble with such a market-driven approach to decision-making is that it undermines the cohesion of the group, which is bound together by shared sacred values. If we think that anything and anyone can be bought, then this cohesion fragments as sacred items lose their special nonmonetary value. For this reason, certain sacred values must exist that cannot be measured by rational analysis. Every society needs things that are taboo and cannot be reduced to trade-offs and comparisons. People do not sign on explicitly to these rules, but we understand that as members of a social group we are expected to share in the same collective sacred values.

Here is the final piece of the puzzle. How can something become sacred? This is where the supersense comes into its own. Society can tell us what is sacred but, to be experienced as sacred, something must become supernatural. It has to be more than mundane. It must possess qualities that are unique and irreplaceable. Discerning such qualities requires a mind designed to sense hidden properties. If something can be copied, duplicated, corrupted, cloned, forged, replaced, or substituted, it is no longer sacred. To arrive at this belief we have to infer that there are hidden supernatural dimensions to our sacred world. And with this thinking comes all the supernatural qualities of connectedness and deeper meaning. We need these to make sense of why we value some things over and above their objective worth. Ironically, it is the supersense that enables us to justify our sacred values. Irrationality makes our beliefs rational because these beliefs hold society together.

AND FINALLY . . .

In this book, I have been sketching an account of how a supersense we all share as members of a highly social species could emerge. Culture and religion simply capitalize on our inclination to infer hidden dimensions to reality. We have discovered that our naturally evolved reasoning mechanisms compel us to make sense of the world by seeking patterns, structures, and mechanisms. We have intuitively done this from the beginning, long before formal education was invented. Supernatural thinking is simply the natural consequence of failing to match our intuitions with the true reality of the world. What's more, these misconceptions are not necessarily discarded over our lifetime. Even as adults, we can simultaneously hold rational models of the world alongside our intuitive notions.

Over the course of childhood, we become participating members of a social group. As young children, we may be the focus of our parents' attention but, as we grow, we must learn to become part of the human race. We must learn to negotiate a social world of competing interests. We must learn to become members of a tribe that shares sacred values.

To achieve this we increasingly become aware of ourselves as unique individuals with unique minds embedded in a society of other unique individuals and minds. We are both individuals and a collective. We see ourselves as part of a group, to be distinguished from other groups. This belief is cemented by our sense that our own group has hidden properties that are essentially different from the invisible properties of other groups.

We mind-read and manipulate others to achieve our individual goals, but we also seek the emotional connections that others provide. We need the totems and sacred objects that bind us together. For many, religion provides these frameworks, but for the rest of us it can be a personal possession, a grubby blanket, a family heirloom, a famous painting, a beautiful statue, a historic monument, a martyr's relic, or a return to the place where we were born. All of our sacred values convey a common sense of connectedness that joins us to each other and to our ancestors. In this way, we are extending ourselves to the rest of humanity from the past to the present.

We may be able to understand the external world through logical cost–benefit analysis, but within each of us is a sacred supersense. If we thought that our partner, spouse, lover, friend, ally, or fellow man did not share these sacred values, we would not trust them and we could not love them. We would see them as fundamentally different from us and even as less human. When people choose to wear a killer's cardigan, they are violating our sacred values and our inherent supersense.

EPILOGUE

Eight months ago on my visit to Gloucester, I discovered that not all buildings associated with evil are levelled to the ground. Fred West's first house in Gloucester, at 25 Midland Road, across a beautiful park from Cromwell Street, still stands today. Somehow this property had escaped the public's attention when it was focused on Cromwell Street. At Midland Road, the dismembered body of his eight-year-old stepdaughter Charmaine was found buried in the cellar. I was unaware of this house until Nick the landlord told me how, despite being a reasonable man, he had felt 'something there' when he visited the property with a view to buying it in 1996. Despite an asking price of only a fraction of the true value, Nick declined. He thought he would have trouble renting it. As it turned out, this is not a problem in a city like Gloucester. It is a deprived area with a large number of migrant workers always in need of affordable accommodation.

On that odd April day, I walked across the park full of people sunning themselves, crossed a busy main road, and found the semidetached property in what was clearly a run-down part of the city. Munchi, a teenage girl, sat on the steps of the house reading a book. I discreetly photographed the house, which immediately made me feel guilty and self-conscious, but I had to ask Munchi about living there. So I approached and tentatively tried to strike up a conversation. I can be an awkward person at the best of times, but I needed to know if she had experienced anything unusual in the house.

Imagine being a teenage girl relaxing with a book on a hot April day and being approached by a middle-aged man wearing an inappropriate leather jacket and asking strange questions. She looked nervous and said that she lived with her cousin, Diana. She was the one to ask. Munchi disappeared inside and returned moments later with Diana, an older woman, who was looking equally suspicious. I asked again, trying to be as relaxed as possible. 'Have you noticed anything strange since you have been living in the house?' Diana was much more open. She said she saw things out of the corner of her eye in the living room. I don't know what I expected to hear. It's such a leading question in the first place. I asked if they knew who Fred West was. Both looked blank and shook their heads.

For a brief instance, I was tempted to tell them the history of their home. How twenty years ago the world's media was focused on Fred and Rosemary West. How people were appalled and disgusted when the details of the gruesome murders of young women and two daughters became known. Telling them this history would have been no stunt with a cardigan to make a point. Munchi and Diana were really living with the past. Their response to this news would be genuine but devastating. What was I to do?

They say ignorance is bliss and to take that away is cruel and unnecessary. So I thanked Munchi and Diana for their time and left them baffled by the strange professor. By the time these words are in print, I expect that Munchi and Diana will have moved on and some other unsuspecting tenants will be living at 25 Midland Road. But if not, Munchi and Diana, I am sorry for not telling you, but I thought it was better for you not to know. There is no essence of evil in your house. It's simply something our minds create. But knowing that doesn't make it feel any more comfortable to be living in the house of a murderer. That's because we are a sacred species.

READER'S NOTES

In Brief

Belief in the supernatural is extremely common in today's modern society. Whether it is religious or secular notions of paranormal activity, most people hold some form of belief that goes beyond the current natural understanding of the world. *SuperSense* attempts to explain this by looking for the origins of such beliefs in children's everyday reasoning. The book surveys the research into early childhood behaviour to reveal that the foundations of many aspects of adult belief appear early in development. This way of thinking is our 'supersense' and, while its influence may disappear with education and increased rational control, it may never entirely go away, especially if the culture supports such beliefs. Moreover, it may become more apparent at times when our ability to exercise rational control is weakened by stress, disease or diminished mental agility. Believing in the supernatural also appears to offer comfort and control when we feel under threat. However, we are not all the same in our reliance on our supersense. There is much room for individual variation. Some of us are more inclined than others towards our supersense, but this may not be a weakness; it may be the basis for why some people are more creative in their thinking. Also our supersense may forge the bonds that hold us together as a society. This is because the supersense may enable individuals to believe and act as if there were some supernatural property that binds them together to form close personal bonds with others. In this way, social cohesion may benefit from this perception of supernatural connection. So, with its natural origins, creative influence and social benefits, it seems unlikely that such a supersense will ever be eradicated entirely by reason.

Brief Biography

Bruce is currently Chair of Developmental Psychology and Director of the Bristol Cognitive Development Centre at the University of Bristol (1999–). He was previously a professor in the Department of Psychology, Harvard University (1995–99) and a visiting scientist at MIT (1994–95). He obtained his first degree in psychology at Dundee University in 1984 and his PhD from Cambridge University in 1991.

About the Author

I was born in Toronto and my middle name is MacFarlane, a legacy of my Scottish heritage from my father's side. My mother is Australian, with the very unusual first name of 'Loyale'. I used to believe for many years that she had two sisters called 'Hope' and 'Faith' but this was just wishful thinking. 'Why Toronto?', I hear you ask. My father was a journalist and plied his art on various continents. By the time, I finally settled in Dundee, Scotland, for the majority of my childhood, I had already lived in Australia, New Zealand and Canada. (If you are wondering, I support Scotland during the Rugby World Cup.) I have an older brother who was also born in Toronto but he doesn't have my mid-Atlantic accent. He is sensible. He is a lawyer. In Dundee, I went to school and then to university where I studied psychology. I then went to Cambridge to conduct research on visual development in babies and completed my PhD in 1991. That year I got married with a 'Dr' in front of my name. My wife is a real doctor and wouldn't marry me until I was doctored. After some post-doc experience at

University College London, we both set off to Boston to sample US academia for a year. By the time we were ready to travel, we were three as my eldest daughter had just been born. When my wife wasn't paying attention, I applied for, and was offered, a professorship at Harvard. What was supposed to be just one year in the US turned into five, by which time we decided that we really did not want to raise our daughter with the same accent as mine. We moved back to Bath, a beautiful city where we never thought we would ever have the opportunity to work. Bristol University, which is not too far away from Bath, offered me a professorial chair in developmental psychology, so I was well pleased. That was ten years ago. We now have a second daughter and we all live in a medieval barn with mice. I still conduct research and teach at Bristol. But I also write books. That's where I am up to now.

HOW DO YOU MEASURE UP ON THE SUPERSENSE?

HOW DO YOU SCORE ON PARANORMAL BELIEF?

In chapter 2, I described the results of Gallup polls surveying the general public in their supernatural beliefs. There are several scales that have been used in research to measure how much belief people have in supernatural phenomena. Listed below are thirteen statements containing opinions designed to quickly measure paranormal beliefs. Following each opinion please indicate how strongly you agree or disagree with the statement, with (1) = Strong Disagreement, (2) = Moderate Disagreement, (3) = Mild Disagreement, (4) = Mild Agreement, (5) = Moderate Agreement and (6) = Strong Agreement. Read the statements carefully as the wording is important.

a) It is probably true that certain people can predict the future quite accurately.
b) For the most part, people who claim to be psychics are in reality just very good actors.
c) It is quite possible for planetary forces to control personality traits.
d) Contrary to scientific opinion, there is some validity to fortune telling.
e) In spite of the laws of science, some people can use their psychic powers to make objects move.
f) As a general rule, any fortune-teller's predictions which come true are just a result of coincidence.
g) Regardless of what you might read in the magazines, people who actually believe in 'magic' rituals are just wasting their time.
h) For the most part, most fortune-tellers' predictions are general and vague. It is just the situation that them believable.
i) In spite of what people think, card readings, for example tarot cards, can reveal a lot about a person and their future.

274

j) Cosmic forces (astrology) can still influence peoples' lives even thought they don't believe in it.

k) Although some people still believe there are people who can actually put a hex on or cast a love spell on some, such belief is only superstition.

l) Contrary to scientific belief, some people can make contact with the dead.

Take your score for items b), e), f), g), and j) then *reverse* the value so that 1=6, 2=5, 3=4, 4=3, 5=2 and 6=1. Now add all your scores together. What was your total?

Research with this scale has shown that the average adult score is 38 in the US and 32 in the UK.

DISCUSSION TOPICS

HAVE YOU HAD ANY SUPERSENSE BELIEFS OR MOMENTS?

If the supersense is within all of us, there should be no embarrassment in talking about our individual supernatural beliefs and behaviours. Why not ask colleagues and friends if they have any interesting super-sense moments and how they came about? Start with the personal sentimental objects, as many of us are happy to describe what they mean to us. If not an object, then it may be a certain place or an event where one experiences a sense of the profound. Our lives are full of such moments of significance though we use different ways to inter-pret them.

WHAT WOULD YOU SAVE FROM A BURNING HOUSE?

In chapter 8, I described how personal possessions become very precious to the individual. Imagine that your house caught fire and you had to save just one item. What would it be and why would you risk your life to save it?

My student Katy Donnelly posed this question to 180 people in an online survey. The top three household items in order were: 1) photo-graphs, 2) jewellery and, 3), their childhood toy. Women rated items they had been given by someone else as more valuable than did men, who valued objects that they had bought themselves as most important. Why do you think that might be the case?

HOW MUCH IS YOUR SUPERSENSE WORTH?

Consider the following tasks and answer honestly if you could do any of them and, if so, for how much: £1; £100; £1000; £100,000;

£1,000,000; or never? You have to fully imagine doing it to get a true insight into your own supersense.

1) Could you drop your most cherished sentimental object into the toilet bowl?
2) Could you wear a murderer's cardigan?
3) Could you stab a copy of a photograph of a loved one through the eyes?

Now consider all three. What order would you rank them in from the least to the worst act? Try this list out with others (maybe as an after-dinner conversation) and see if you all agree. I bet you will not.

IS YOUR SUPERSENSE STRONG?

The 'magical ideation' scale that I talked about in chapter 9 is a measure of one's tendency to the supersense. It was devised by Mark Eckblad and Loren Chapman from the University of Wisonsin-Madison in 1983 and has been used to look at magical (supernatural) thinking in the general public. How do you score on this? Read each item and tick the true or false box.

1	Some people can make me aware of them just by thinking about me.	False	True
2	I have had the momentary feeling that I might not be human.	False	True
3	I have sometimes been fearful of stepping on pavement cracks.	False	True
4	I think I could learn to read other peoples' minds if I wanted to.	False	True
5	Horoscopes are right too often for it to be a coincidence.	False	True
6	Things sometimes seem to be in different places when I get home, even though no one has been there.	False	True
7	Numbers like 13 and 7 have no special power.	True	False
8	I have occasionally had the silly feeling that a TV or radio broadcaster knew I was listening to them.	False	True
9	I have worried that other people on other planets may be influencing what is happening on earth	False	True
10	The government refuses to tell the truth about flying saucers.	False	True
11	I have felt that there were messages for me in the way things are arranged, like in a store window.	False	True
12	I have never doubted that my dreams are the products of my own mind.	True	False
13	Good luck charms don't work.	True	False
14	I have noticed sounds on my records that are not there at other times.	False	True
15	The hand motions that strangers make seem to influence me at times.	False	True
16	I almost never dream about things before they happen.	True	False
17	I have had the momentary feeling that someone's place has been taken by a look-alike.	False	True
18	It is not possible to harm others merely by thinking bad thoughts about them.	True	False
19	I have sometimes sensed an evil presence around me, although I could not see it.	False	True
20	I sometimes have a feeling of gaining or losing energy when certain people look at me or touch me.	False	True
21	I have sometimes had the passing thought that strangers are in love with me.	False	True
22	I have never had the feeling that certain thoughts of mine really belonged to someone else.	True	False
23	When introduced to strangers, I rarely wonder whether I have known them before.	True	False
24	If reincarnation were true, it would explain some unusual experiences I have had.	False	True
25	People often behave so strangely that one wonders if they are part of an experiment.	False	True
26	At times I perform certain little rituals to ward off negative influences.	False	True
27	I have felt that I might cause something to happen just by thinking about it.	False	True
28	I have wondered whether the spirits of the dead can influence the living.	False	True
29	At times I have felt that a professor's lecture was meant especially for me.	False	True
30	I have sometimes felt that strangers were reading my mind.	False	True

Now add up all the boxes in the right-hand column that you ticked. The average score (based on 1,500 US students) was 9 for males and 10 for females. How did you score? Do you think the questions measure supernatural thinking?

WHAT PATTERNS DO YOU SEE?

In the book, I talk about how some of us with a strong supersense are more inclined towards seeing structure and patterns in the world. Test your self with the following set of images taken from the 'Snowy Pictures task' (© ETS) used to assess this capacity. Can you see any hidden patterns? Answers are at the bottom.

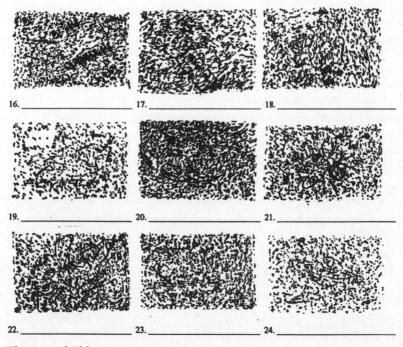

16. _____ 17. _____ 18. _____

19. _____ 20. _____ 21. _____

22. _____ 23. _____ 24. _____

There are hidden patterns in images, 17, 19, 21, 22 & 24.

ACKNOWLEDGEMENTS

Many of the ideas I am presenting are based on academic research but they have been road-tested in public lectures and informal gatherings throughout the UK. When you need to talk in a way that most people can understand, it forces you to cut to the chase: to stop waffling and get to the heart of the matter. Academics are trained to recognize weaknesses, but in doing so we can become preoccupied with hedging our bets, qualifying our assertions and being over-cautious in our interpretations, so that we often seem to be sitting on a fence of indecision. That's why writing a book such as this one might make an academic nervous.

Thankfully, I have been helped along the way. I am indebted to colleagues, students, friends, and family who have read different versions, made suggestions and generally encouraged me to be less nervous about the book. In alphabetical order I would like to thank Sara Baker, Horace Barlow, Susan Blackmore, Paul Bloom, Peter Brugger, Zoltan Dienes, Katy Donnelly, Shiri Einav, Norman Freeman, Susan Gelman, Iain Gilchrist, Thalia Gjersoe, Richard Gregory, Charlotte Hardie, Hilary and Peter Hodgson, Loyale Hood, Alison and Ross Hood, Marjaana Lindeman, Neil Macrae, Peter Millican, Steven Pinker, Paul Rozin, Reba Rosenberg, Ali Smith, Elaine Snell, Andrew Stuart, Arno van Voorst, Alice Wilson and Stephen Woolley. I also want to thank all those working in the Bristol Cognitive Development Centre who have supported me in this project.

Bruce Hood

The thesis I present in this book is one that I have been developing over the course of my professional academic life. However, it really came to life after my arrival at the University of Bristol, which has proved a supportive and nurturing environment in which to work. I could not have maintained my research programme without the support of the UK agencies that fund my work. I would like to thank the Economic and Social Research Council and the Medical Research Council. In addition, I would also like to thank the private foundations that support me including the Leverhulme Trust (UK), the Esmée Fairbairn Foundation (UK) and the Bial Foundation (Portugal).

For this UK edition, I am particularly grateful to the support, encouragement, and vision, of Andreas Campomar of Constable & Robinson.

I dedicate this book to my family.

SOURCE NOTES

PROLOGUE

1. Sean Coughlan, 'What Happens to the Houses of Horror?', *BBC News*, 5 April, 2004, available at: http://news.bbc.co.uk/1/hi/ magazine/3593137.stm.

2. Although the National Association of Realtors in the United States requires its members to reveal all physical factors that might affect the desirability of a house, there is no consensus when it comes to the psychological factors that may stigmatize a home.

3. 'Couple Lose House of Horror Case', *BBC News*, 7 February, 2004, available at: http://news.bbc.co.uk/1/hi/england/west_yorkshire/ 3492936.stm

4. There were several reports of the security surrounding the demolition; see 'Soham Murder House Is Demolished', *BBC News*, 3 April, 2004, available at: http://news.bbc.co.uk/1/low/england/ cambridgeshire/3595801.stm; and Tony Thompson, 'As Day Breaks, Huntley's House Is Turned into Dust and Rubble', the *Guardian*, 4 April, 2004, available at: http://observer.guardian.co.uk/uk_news/ story/0,,1185348,00.html.

5. The outcome of the celebrity memorabilia auction at Christie's is reported online at: http://www.telegraph.co.uk/news/newstopics/ celebritynews/2282365/Beatles-Sgt-Pepper-drumskin-in-record-sale.html

6. The designers of Princess Diana's wedding dress, David and Elizabeth Emanuel, are currently selling – for £1,000 – a book with a swatch of the wedding dress that she wore during fittings; available from http://www.adressfordiana.com.

7. The word 'fetish' (from the Latin *facticious*, for 'artificial') was originally coined by Charles de Brosses in 1757 to refer to objects believed by West African tribes to have supernatural powers.

8. James Randi discusses the cardigan stunt and tells readers about his own experience with Brother André's relic at: http://www.randi.org/jr/2006–09/092206bad.html.

9. For example, the Pan Fest, a pagan festival held on the prairies in Alberta, Canada, during Lammas in August.

10. Tony Blair's superstition about his shoes is reported in *The Times* online at http://www.timesonline.co.uk/tol/news/politics/the _ blair _ years/article1969242.ece

11. J. Curl. 'McCain channeling as his luck toward 2008 race; Keeps assortment of talismans to ward off a Democratic win.' *The Washington Times,* 16 April 2008.

12. I am indebted to Steven Pinker, who introduced me to Philip Tetlock's work on sacred values that seeded the idea in my head that a supernatural sense makes such beliefs so powerful.

13. http://www.happynews.com/news/5132008/fans-long-ashes-scattered-sporting-sites.htm

CHAPTER ONE

1. P. Le Loyer, introduction to *IIII Livres des Spectres, ou, Apparitions et Visions d'Espirits, Anges, et Demons se monstrans sensiblement aux hommes*, 2nd edn, translated by Zachery Jones (British Library, 1605).

2. www.ted.com/talks/view/id/22.

3. D. Clarke, 'Experience and Other Reasons Given for Belief and Disbelief in the Paranormal and Religious Phenomena', *Journal of the Society for Psychical Research* 60 (1995): 371–84.

4. A. Grimby, 'Bereavement Among Elderly People: Grief Reactions, Post-Bereavement Hallucinations and Quality of Life', *Acta Psychiatrica Scandinavia* 87 (1993): 72–80.

5. This probability is based on any two people sharing a birthday. Conversely, if you were asked how many people would need to be at a party for you to meet someone who shares your birthday at half the parties you attend, the number seems unreasonably high at 253. Those of you unconvinced by these figures can check out Ian Stewart, *The*

Magical Maze: Seeing the World Through Mathematical Eyes (Weidenfeld & Nicholson, 1997). Peter Milligan from Oxford University told me about the soccer example.

6. M. Plimmer and B. King, *Beyond Coincidence* (Icon Books, 2005), p. 4.

7. W. James, *On Varieties of Religious Experiences* (Basic Books, 1902), p. 58.

8. Ibid., p. 510.

9. S. Vyse, *Believing in Magic: The Psychology of Superstition* (Oxford University Press, 1997) p. 60.

10. Many Christian holidays such as Christmas and Easter incorporate elements from earlier pagan ceremonies. For example, Yule logs can be traced back to Norse pagan festivals, in which they were symbols of health and productivity. Mistletoe was also used in Norse pagan ceremonies and is linked to fertility by the resemblance of the fruit's content to semen.

11. R. Dawkins, *Unweaving the Rainbow* (Penguin Books, 2006), p. 36.

12. N. Chomsky, *Syntactic Structures* (Mouton, 1957), p. 15.

13. The Nobel physicist Richard Feynman once remarked that he found it easier to picture invisible angels than light rays; quoted in A. Lightman, *A Sense of the Mysterious: Science and the Human Spirit* (Vintage Books, 2005): 'Physics has galloped off into territories where our bodies cannot go' (p. 63).

14. This phrase was first coined by the psychologists Leda Cosmides and John Tooby in 'Origins of Domain Specificity: The Evolution of Functional Organization', in *Mapping the Mind: Domain Specificity in Cognition and Culture*, edited by L. A. Hirschfeld and S. A. Gelman (Cambridge University Press, 1994).

15. The neurophysiologist Rudiger von der Heydt of Johns Hopkins University demonstrated the presence of 'end-stopped cells' in the visual areas of the brain that are activated by such patterns as if the illusory contour were really there.

16. H. Ghim, 'Evidence for Perceptual Organization in Infants: Perception of Subjective Contours by Young Infants', *Infant Behaviour and Development* 13 (1990): 221–48.

17. The website for the World Rock Paper Scissors Society can be found at http://www.worldrps.com.

18. A. D. Baddeley, 'The Capacity for Generating Information by Randomization', *Quarterly Journal of Experimental Psychology* 18 (1966): 119–29.

19. A. M. Leslie, 'Spatiotemporal Continuity and Perception of Casuality in Infants', *Perception* 13 (1984): 287–305.

20. Of all the sports, tennis seems to produce the greatest share of superstitious rituals in both men and women. Like John McEnroe, Martina Hingis also would never step on the lines between points. Marat Safin travels with an 'evil eye' given to him by his sister to help ward off malevolent stares. Goran Ivanisevic follows a strict pregame regimen: eating at the same table of the same restaurant and ordering a set feast of fish soup, lamb, and ice cream with chocolate sauce.

21. *The Young Ones*, Episode no. 12 ('Summer Holiday'), first broadcast 19 June, 1984 by BBC2. Directed by Geoff Posner and written by Ben Elton, Rik Mayall and Lise Mayer.

22. Tim Lovejoy, interview with David Beckham, broadcast on the United Kingdom's ITV1, 2006.

23. E. J. Langer, 'The Illusion of Control', *Journal of Personality and Social Psychology* 32 (1975): 311–28.

24. G. Keinan, 'The Effects of Stress and Desire for Control on Superstitious Behaviour', *Personality and Social Psychology Bulletin* 28, (2002): 102–8.

25. T. V. Salomons, T. Johnstone, M. Backonja, and R. J. Davidson, 'Perceived Controllability Modulates the Neural Response to Pain', *Journal of Neuroscience* 24 (2004): 7199–203.

26. J. A. Whitson and A. D. Galinsky, 'Lacking Control Increases Illusory Pattern Perception', *Science*, 322 (2008): 115–17.

27. E. Pronin, D. M. Wegner, K. McCarthy, and S. Rodriguez, 'Everyday Magical Powers: The Role of Apparent Mental Causation in the Overestimation of Personal Influence', *Journal of Personality and Social Psychology* 91 (2006): 218–31.

28. Another famous urban myth is that Galileo dropped cannonballs of different weights from the Leaning Tower of Pisa to demonstrate that they would land at the same time. In fact, others, such as the Flemish engineer Simon Stevin, had already published the results of experiments in 1586 on falling weights before Galileo's appointment as professor of mathematics at Pisa in 1612.

29. A. B. Champagne, L. E. Klopfer, and J. H. Anderson, 'Factors Influencing the Learning of Classical Mechanics', *American Journal of Physics* 48 (1980): 1074–9.

CHAPTER TWO

1. Elli Leadbeater, 'Woolly Ruse Incites Irrationality', 4 September, 2006, *BBC News*, available at: http://news.bbc.co.uk/1/hi/sci/tech/5314164.stm.
2. M. Van Vugt and C. M. Hart, 'Social Identity as Social Glue: The Origins of Group Loyalty', *Journal of Personality and Social Psychology* 86 (2004): 585–98.
3. G. Le Bon, *The Crowd: A Study of the Popular Mind* (1896; reprint, Ernst Benn Ltd/Transaction Publishers, 1995), p. 148.
4. N. Ambady and R. Rosenthal, 'Thin Slices of Expressive Behaviour as Predictors of Interpersonal Consequences: A Meta-analysis', *Psychological Bulletin* 111 (1992): 256–74; N. Ambady and R. Rosenthal, 'Half a Minute: Predicting Teacher Evaluations from Thin Slices of Nonverbal Behaviour and Physical Attractiveness', *Journal of Personality and Social Psychology* 64 (1993): 431–41.
5. A. Damasio, *Descartes' Error* (Vintage Books, 1994).
6. D. C. Fowles, 'The Three Arousal Model: Implications for Fray's Two-Factor Learning Theory for Heart Rate, Electrodermal Activity, and Psychopathy', *Psychophysiology* 17 (1980): 87–104.
7. P. Rozin, M. Markwith, and C. Nemeroff, 'Magical Contagion Beliefs and Fear of AIDS', *Journal of Applied Social Psychology* 22 (1992): 1081–92.
8. The Festival del Burro, or Festival of the Donkey, takes place in March in the town of San Antero near Cordova in northern Colombia.
9. A. Silverman, 'Sexton Admits 2000 Killing of Atsuko Ikeda. July 29th 2006.' Winooski police press release: www.winooskipolice.com/Press%20Release/Sexton.htm
10. C. Zhong and K. Liljenquist, 'Washing Away Your Sins: Threatened Morality and Physical Cleansing', *Science* 313 (2006): 1451–2.
11. *The Exorcist*, directed by William Friedkin (Hoya Productions, 1973).
12. R. Wiseman, *Quirkology: The Curious Science of Everyday Lives* (Macmillan, 2007).
13. Available from the Gallup Organization, Princeton, N.J.: http://www.gallup.com.
14. http://www.gallup.com/poll/19558/Paranormal-Beliefs-Come-SuperNaturally-Some.aspx

CHAPTER THREE

1. A. Forbes and T. R. Crowder, 'The Problem of Franco-Cantabrian Abstract Signs: Agenda for a New Approach', *World Archaeology* 10 (1979): 350–66.

2. D. Lewis-Williams, *The Mind in the Cave: Consciousness and the Origins of Art* (Thames & Hudson, 2004).

3. Gallup poll, May 2007, available from the Gallup Organization, Princeton, N.J., http://www.gallup.com.

4. J. Sulston, 'Why won't the public put their faith in scientists?' *THES (Times Higher Education Supplement)*, 9 June 2005 pp.18-19.

5. N. Humphrey, *Leaps of Faith: Science, Miracles, and the Search for Supernatural Consolation* (Springer, 1999), p. 8.

6. T. Hobbes, *Leviathan* (1651; reprint, W. W. Norton, 1996).

7. This interview is available at BBC Radio 4, 'Science: The Material World', http://www.bbc.co.uk/radio4/science/thematerialworld_20060921.shtml.

8. R. Dawkins, *The God Delusion* (Bantam Press, 2006), p. 36.

9. J. Barrett, *Why Would Anyone Believe in God?* (AltaMira Press, 2004).

10. R. Baillargeon, J. DeVos, and M. Graber, 'Location Memory in Eight-Month-Old Infants in a Non-Search AB Task: Further Evidence', *Cognitive Development* 4 (1989): 345–67.

11. J. Connellan, S. Baron-Cohen, S. Wheelwright, A. Batki, and J. Ahluwalia, 'Sex Differences in Human Neonatal Social Perception', *Infant Behaviour and Development* 23 (2000): 113–18.

12. G. Huntley-Fenner, S. Carey, and A. Solimando, 'Objects Are Individuals but Stuff Doesn't Count: Perceived Rigidity and Cohesiveness Influence Infants' Representations of Small Groups of Distinct Entities', *Cognition* 85 (2002): 203–21.

13. R. Baillargeon, A. Needham, and J. DeVos, 'The Development of Young Infants' Intuitions About Support', *Early Development and Parenting* 1 (1992): 69–78.

14. A. Shtulman and S. Carey, 'Improbable or Impossible? How Children Reason About the Possibility of Extraordinary Events', *Child Development* 78 (2007): 1015–32.

15. D. C. Dennett, *Breaking the Spell: Religion as a Natural Phenomenon* (Allen Lane, 2005).

16. A. Tversky and D. Kahneman, 'Extension Versus Intuitive Reasoning:

The Conjunction Fallacy in Probability Judgement', *Psychological Review* 90 (1983): 293–315.

17. For details of the survey, see Zoological Society of London, 'Nation's Phobias Revealed', 27 October, 2005, available at: http://www.zsl.org/info/media/press-releases/null,1780, PR.html.

18. J. B. Watson and R. Raynor, 'Conditioned Emotional Reactions', *Journal of Experimental Psychology* 3 (1920): 1–14.

19. M. E. P. Seligman, 'Phobias and Preparedness', *Behaviour Therapy* 2 (1971): 307–20. In reviewing all the experimental data, Rich McNally concluded that, while various aspects of Seligman's theory are questionable, his one unquestionable assertion is that 'most phobias are associated with threats of evolutionary significance'. R. McNally, 'Preparedness and Phobias: A Review', *Psychological Bulletin* 101 (1987): 283–303.

20. S. Atran, *In Gods We Trust: The Evolutionary Landscape of Religion* (Oxford University Press, 2002).

21. P. Boyer, *Religion Explained: The Human Instincts That Fashion Gods, Spirits, and Ancestors* (William Heinemann, 2001).

22. R. Dawkins, *The Blind Watchmaker* (Penguin Books, 1986), p. 316.

23. L. Rozenblit and F. C. Keil, 'The Misunderstood Limits of Folk Science: An Illusion of Explanatory Depth', *Cognitive Science* 26 (2002): 521–62.

24. H. Spencer, *Principles of Biology* (Williams & Norgate, 1864).

25. E. M. Evans, 'Conceptual Change and Evolutionary Biology: A Developmental Analysis', in *Handbook of Research on Conceptual Change*, edited by S. Vosniadou (Taylor & Francis Group, 2008).

26. The majority of reef fish change sex at some point in their life, and in fact, those that do not are in the minority (source: Aaron Rice, Davidson College).

27. Human embryos start out as female and, in the absence of a Y chromosome, continue to develop as female.

28. Cladistics is the science of mapping the comparative genetic code of all living things to trace the tree of life. For an accessible introduction, read S. Jones, *Almost Like a Whale* (Doubleday, 1999).

29. For an extensive web resource on creation myths, try 'Magic Tails', available at: www.magictails.com/creationlinks.html.

30. E.M. Evans, 'The Emergence of Beliefs About the Origins of Species in School-Age Children', *Merrill-Palmer Quarterly: A Journal of Developmental Psychology* 46 (2000): 221–54.

31. Cited in Dawkins, *The God Delusion*, p. 102.

32. Carnegie Commission, *National Survey of Higher Education: Faculty Study* (McGraw-Hill, 1969).

33. Following his recent recovery, Dennett thanked friends who prayed for him: 'Thanks, I appreciate it, but did you also sacrifice a goat?'; see www.edge.org/3rd_culture/dennett06/dennett06_index.html.

34. E. H. Ecklund and C. P. Scheitle, 'Religion Among Academic Scientists: Distinctions, Disciplines, and Demographics', *Social Problems* 54 (2007): 289–307.

35. Jan Walsh, *Living TV Paranormal Report* (Consumer Analysis Group, 2002).

36. S. Harris, *The End of Faith: Religion, Terror, and the Future of Reason* (W. W. Norton, 2004).

37. Meera Nanda has been one of the more eloquent critics of Sam Harris; see 'Trading Faith for Spirituality: The Mystifications of Sam Harris' posted 16 December, 2006, http://www.sacw.net/free/Trading%20Faith%20for%20Spirituality_%20The%20Mystifications%20of%20Sam%20Harris.html.

38. Dennett, *Breaking the Spell* p. 21.

39. P. Zuckerman, 'Atheism: Contemporary Rates and Patterns', in *Cambridge Companion to Atheism*, edited by M. Martin (Cambridge University Press, 2005).

40. E. H. Lenneberg, *Biological Foundations of Language* (Wiley, 1967).

41. D. S. Lundsay, P. C. Jack, and M. A. Christian, 'Other-Race Perception', *Journal of Applied Psychology* 76 (1991): 587–9.

42. After only about twelve hours of accumulated exposure to their own mother's face: newborns show a preference for her face compared to other mothers: I. W. R. Bushnell, 'The Origins of Face Perception', in *The Development of Sensory, Motor, and Cognitive Capacities in Early Infancy: From Perception to Cognition*, edited by F. Simion and G. Butterworth. (Psychology Press/Hove, 1998).

43. D. J. Kelly, P. C. Quinn, A. M. Slater, K. Lee, L. Ge, and O. Pascalis, 'The Other-Race Effect Develops During Infancy', *Psychological Science* 18 (2007): 1084–9.

44. The interview with Peter and Christopher Hitchens can be found at http://www.bbc.co.uk/radio4/today/listenagain/ram/today4_20070619.ram.

45. T. J. Bouchard Jr., M. McGue, D. Lykken, and A. Tellegen, 'Intrinsic and Extrinsic Religiousness: Genetic and Environmental Influences

and Personality Correlates', *Twin Research* 2 (1999): 88–98.

46. K. M. Kirk, L. J. Eaves, and N. G. Martin, 'Self-transcendence as a Measure of Spirituality in a Sample of Older Australian Twins', *Twin Research* 2 (1999): 81–7; L. B. Koenig, M. McGue, R. F. Krueger, and T. J. Bouchard Jr., 'Genetic and Environmental Influences on Religiousness: Findings for Retrospective and Current Religiousness Ratings', *Journal of Personality* 73 (2005): 471–88.

47. D. Hamer, *The God Gene: How Faith Is Hardwired into Our Genes* (Doubleday, 2004).

48. A. Newberg, E. D'Aquili, and V. Rause, *Why God Won't Go Away: Brain Science and the Biology of Belief* (Ballantine Books, 2001).

49. Isaac Bashevis Singer, quoted by Stefan Kanfer in 'Isaac Singer's Promised City', *City Journal*, Summer, 1997, http://www.city-journal.org/html/7_3_urbanities-issac.html.

50. M. Hutson, 'Magical Thinking: Even Hard-core Sceptics Can't Help but Find Sympathy in the Fabric of the Universe', *Psychology Today* (March–April 2008). I e-mailed Lori Blanc, and she confirmed the reports in the press.

51. D. Gilbert, *Stumbling on Happiness* (HarperCollins, 2006).

CHAPTER FOUR

1. W. James, *Principles of Psychology* (1890; reprint, Harvard University Press, 1983).

2. See also J. B. Watson, *Behaviourism*, rev. edn (University of Chicago Press, 1930).

3. A. Jolly, *Lucy's Legacy: Sex and Intelligence in Human Evolution* (Harvard University Press, 1999).

4. I am indebted to the neuropathologist Seth Love for confirming that there is reactivation of infantile reflexes following brain damage.

5. J. Atkinson, B. Hood, J. Wattam-Bell, S. Anker, and J. Tricklebank, 'Development of Orientation Discrimination in Infancy', *Perception* 17 (1988), 587–95.

6. A. J. DeCasper and M. J. Spence, 'Prenatal Maternal Speech Influences Newborns' Perception of Speech Sounds', *Infant Behaviour and Development* 9 (1986): 133–50.

7. P. G. Hepper, 'Fetal "Soap" Addiction', *The Lancet* (11 June, 1988); 1347–8.

8. V. Reddy, 'Playing with Others' Expectations: Teasing and Mucking About in the First Year', in *Natural Theories of Mind*, edited by A. Whiten (Oxford University Press, 1991).

9. F. J. Zimmerman, D. A. Christakis, and A. N. Meltzoff, 'Associations Between Media Viewing and Language Development in Children Under Age Two Years', *Journal of Pediatrics* (online press release, 7 August, 2007). The Walt Disney Company has demanded that the University of Washington, where the study was conducted, retract the press release. The University of Washington has stood behind the press release.http://www.washington.edu/alumni/uwnews/links/200709/videos.html

10. The 'Mozart effect' is the claim popularized by Don Campbell in his 1997 book *The Mozart Effect: Tapping the Power of Music to Heal the Body, Strengthen the Mind, and Unlock the Creative Spirit* that listening to classical music increases your IQ. Such was the power of this disputed claim that Zell Miller, the governor of Georgia, announced that his proposed state budget would include $105,000 a year to provide every child born in Georgia with a tape or CD of classical music. To make his point, Miller played legislators some of Beethoven's 'Ode to Joy' on a tape recorder and asked, 'Now, don't you feel smarter already?'

11. The Wimmer Ferguson Infant Stim-Mobile is the black-and-white-patterned toy that has found its way into many a home, including ours. The principle behind it is valid. In the first months of life, babies are attracted to high-contrast features in the visual world, but those features don't have to be black and white. Any area of brightness and darkness attracts their attention, such as overhead lighting, the dark curtains against a sunlit window, or your hairline if you are a brunette. When I worked on visual development, many brunette mothers used to ask me why their newborns never seemed to look them straight in the eye.

12. J. T. Bruer, *The Myth of the First Three Years: A New Understanding of Early Brain Development and Lifelong Learning* (Free Press, 1999).

13. 'Study Reveals: Babies Are Stupid', *The Onion* (1999), available at: http://www.onion.demon.co.uk/theonion/other/babies/stupidbabies.htm. Check out some very cute babies being made fun of.

14. 'Babies Are Smarter Than You Think', *Life* (July 1993).

15. Minsky quoted in Sherry Turkle, *Life on the Screen: Identity in the Age of the Internet* (Simon & Schuster, 1997), p. 137: 'The mind is a meat machine.'

16. The story can be found all over the Internet, but I believe the most sensible consideration of the topic can be found in J. Hutchins, 'The Whiskey Was Invisible: Or, Persistent Myths of MT', *MT News International* 11 (1995), 17–18.

17. J. Locke, *An Essay Concerning Human Understanding* (1690; reprint, E. P. Dutton, 1947).

18. R. Descartes, *Meditations on First Philosophy*, translated by J. Veitch (1647; reprint, Prometheus Books, 1901); I. Kant, *Critique of Pure Reason*, translated by N. K. Smith (1781; reprint, St Martin's Press, 1965).

19. E. S. Spelke, 'Principles of Object Perception', *Cognitive Science* 14 (1990): 29–56.

20. J. B. Watson, *Behaviourism* (University of Chicago Press, 1930), p. 104.

21. B. F. Skinner, 'Superstition in the Pigeon', *Journal of Experimental Psychology* 38 (1948): 168–72.

22. The baby crib was likened to the 'Skinner Boxes' that Skinner had developed for the experimental studies of the effects of rewards on animal behaviour; L. Slater, *Opening Skinner's Box: Great Psychological Experiments of the Twentieth Century* (Bill Daniels Co., 2004).

23. In the *Ladies' Home Journal* article (October 1945), Skinner described the benefits of raising a child in a thermostatically controlled environment so that the baby only needed to wear a diaper. He noted that behaviour and health seemed to thrive in the Air-Crib. An independent questionnaire evaluation by John M. Gray sent to 73 couples who raised 130 babies in the Air-Crib confirmed Skinner's remarkable claims. All but three of these couples described the device as 'wonderful'. Following the slur in *Opening Skinner's Box*, Deborah Skinner wrote a scathing response to the book, 'I Was Not a Lab Rat', the *Guardian*, 12 March, 2004.

24. H. Gardner, *The Mind's New Science: A History of the Cognitive Revolution* (Basic Books, 1985).

25. This scenario is a philosophical issue described as 'the brain in a vat' by Hilary Putnam in chapter 1 of *Reason, Truth, and History* (Cambridge University Press, 1982), pp. 1–21.

26. C. von Hofsten, 'Development of Visually Guided Reaching: The Approach Phase', *Journal of Human Movement Studies* 5 (1979): 160–78.

27. J. Piaget, *The Construction of Reality in the Child* (Basic Books, 1954).

28. There have been literally hundreds of infant studies based on the principle of the magic trick, but the most famous is probably one of

the first involving a solid block that appears to pass through another solid object. R. Baillargeon, E. S. Spelke, and S. Wasserman, 'Object Permanence in Five-Month-Old Infants', *Cognition* 20 (1985), 191–208.

29. K. Wynn, 'Addition and Subtraction by Human Infants', *Nature* 358 (1992): 749–50.

30. E. S. Spelke, 'Core Knowledge', *American Psychologist* 55 (2000): 1233–43.

31. D. Poulon-Dubois, 'Infants' Distinctions Between Animate and Inanimate Objects: The Origins of Naive Psychology', in *Early Social Cognition: Understanding Others in the First Months of Life*, edited by P. Rochat (Erlbaum, 1999).

32. A. L. Woodward, 'Infants Selectively Encode the Goal Object of an Actor's Reach', *Cognition* 69 (1998): 1–34; see also V. Kuhlmeier, K. Wynn, and P. Bloom, 'Attribution of Dispositional States by Twelve-Month-Old Infants', *Psychological Science* 14 (2003): 402–8.

33. A. Karmiloff-Smith, *Beyond Modularity: A Developmental Perspective on Cognitive Science* (MIT Press, 1992).

34. G. L. Murphy and D. L. Medin, 'The Role of Theories in Conceptual Coherence', *Psychological Review* 3 (1985): 289–316.

35. A. Karmiloff-Smith, B. In helder, 'If You Want to Get Ahead, Get a Theory', *Cognition* 23 (1975): 95–147.

36. B. M. Hood, 'Gravity Rules for Two- to Four-Year-Olds?' *Cognitive Development* 10 (1995): 577–98.

37. M. Tomonaga, T. Imura, Y. Mizuno, and M. Tanaka, 'Gravity Bias in Young and Adult Chimpanzees (*Pan troglodytes*): Tests with a Modified Opaque-Tubes Task', *Developmental Science* 10 (2007): 411–21; see also B. Osthaus, A. M. Slater, and S. E. G. Lea, 'Can Dogs Defy Gravity? A Comparison with the Human Infant and Nonhuman Primate', *Developmental Science* 6 (2003): 489–97.

38. I. K. Kim and E. S. Spelke, 'Perception and Understanding of Effects of Gravity and Inertia on Object Motion', *Developmental Science* 2 (1999): 339–62.

39. M. K. Kaiser, D. R. Proffitt, and M. McCloskey, 'The Development of Beliefs About Falling Objects', *Perception and Psychophysics* 38 (1985): 533–9.

40. M. McCloskey, A. Washburn, and L. Felch, 'Intuitive Physics: The

Straight-Down Belief and Its Origin', *Journal of Experimental Psychology: Learning, Memory, and Cognition* 9 (1983): 636–49.

41. J. Piaget, *The Child's Conception of the World* (Routledge & Kegan Paul, 1929).

42. D. Kelemen, 'The Scope of Teleological Thinking in Preschool Children', *Cognition* 70 (1999): 241–72.

43. D. Kelemen, 'Are Children "Intuitive Theists"?' *Psychological Science* 15 (2004): 295–301.

44. J. Piaget. *The Child's Conception of the World* (Routledge & Kegan Paul, 1929).

45. D. Hume, *Natural History of Religion* (1757, reprint Clarendon Press, 1976).

46. J. D. Woolley, 'Thinking About Fantasy: Are Children Fundamentally Different Thinkers and Believers from Adults?' *Child Development* 68 (1997): 991–1011; J. D. Woolley and K. E. Phelps, 'Young Children's Practical Reasoning About Imagination', *British Journal of Developmental Psychology* 12 (1994): 53–67.

47. C. N. Johnson and P. L. Harris, 'Magic: Special but Not Excluded', *British Journal of Developmental Psychology* 12 (1994): 35–51.

48. E. V. Subbotsky, 'Explanations of Unusual Events: Phenomenalistic Causal Judgements in Children and Adults', *British Journal of Developmental Psychology* 15 (1997): 13–36.

49. J. Haidt, F. Bjorkland, and S. Murphy, 'Moral Dumbfounding: When Intuitions Finds No Reason', unpublished study (10 August, 2000).

50. 'Clarke's third law', in A. C. Clarke, *Profiles of the Future: An Inquiry into the Limits of the Possible* (Harper & Row, 1962).

51. M. Mead, 'An Investigation of the Thought of Primitive Children with Special Reference to Animism', *Journal of the Royal Anthropological Institute* 62 (1932): 173–90.

52. G. Bennett, *Traditions of Belief: Women, Folklore, and the Supernatural Today* (Pelican Books, 1987).

53. J. Pole, N. Berenson, D. Sass, D. Young, and T. Blass, 'Walking Under a Ladder: A Field Experiment on Superstitious Behaviour', *Personality and Social Psychology Bulletin* 1 (1974): 10–12.

54. J. M. Bering, 'The Folk Psychology of Souls,' *Behavioural and Brain Sciences* 29 (2006): 453–98.

55. Johnson and Harris, 'Magic: Special but Not Excluded', *British Journal of Developmental Psychology* 12 (1994): 35–51

56. E. V. Subbotsky, 'Early Rationality and Magical Thinking in Preschoolers: Space and Time', *British Journal of Developmental Psychology* 12 (1994): 97–108.

57. I. Opie and P. Opie, *The Lore and Language of School Children* (Oxford University Press, 1959), p. 210.

58. N. Humphrey, *Consciousness Regained* (Oxford University Press, 1984).

CHAPTER FIVE

1. S. Baron-Cohen, *The Essential Difference: The Truth About the Male and Female Brain* (Basic Books, 2005).

2. D. J. Povinelli and T. J. Eddy, 'What Young Chimpanzees Know About Seeing', *Monographs of the Society for Research in Child Development* 61, no. 2, serial no. 247 (1996).

3. K. Lorenz, 'Part and Parcel in Animal and Human Societies', in *Studies in Animal and Human Behaviour*, vol. 2, edited by K. Lorenz (Harvard University Press, 1971).

4. S. Goldberg, S. L. Blumberg, and A. Kriger, 'Menarche and Interest in Infants: Biological and Social Influences', *Child Development* 53 (1982): 1544–50.

5. M. H. Johnson, S. Dziurawiec, H. Ellis, and J. Morton, 'Newborns' Preferential Tracking for Face-like Stimuli and Its Subsequent Decline', *Cognition* 40 (1991): 1–19.

6. M. H. Johnson, 'Imprinting and the Development of Face Recognition: From Chick to Man', *Current Directions in Psychological Science* 1 (1992): 52–5.

7. N. Kanwisher, J. McDermott, and M. Chun, 'The Fusiform Face Area: A Module in Human Extrastriate Cortex Specialized for the Perception of Faces', *Journal of Neuroscience* 17 (1997): 4302–11. Actually, there is now some dispute whether the area is specific to faces or any special category of well-known objects. Given that faces are the most common diverse objects we encounter, this suggests that the area probably evolved primarily for faces.

8. O. Sacks, *The Man Who Mistook His Wife for a Hat* (Pan Books, 1998).

9. J. R. Harding, 'The Case of the Haunted Scrotum', *Journal of the Royal Society of Medicine* 89 (1996): 600.

10. S. Guthrie, *Faces in the Clouds: A New Theory of Religion* (Oxford University Press, 1993).

11. '"Virgin Mary" Toast Fetches $28,000', *BBC News*, 23 November, 2004, available at: http://news.bbc.co.uk/1/hi/world/americas/4034787.stm. 'Woman Sees Face of Jesus in Ultrasound Photo', WKYC.com, 11 April, 2005, available at: http://www.wkyc.com/news/news_fullstory.asp?id=33156.

12. Z. Wang and W. Z. Aragona, 'Neurochemical Regulation of Pair Bonding in Male Prairie Voles', *Physiology and Behaviour* 83 (2004): 319–28.

13. G. Johansson, 'Visual Perception of Biological Motion and a Model for Its Analysis', *Perception and Psychophysics* 14 (1973): 201–11.

14. B. I. Bertenthal, 'Perception of Biomechanical Motions by Infants: Intrinsic Image and Knowledge-Based Constraints', in *Carnegie Symposium on Cognition: Visual Perception and Cognition in Infancy*, edited by C. Granrud (Erlbaum, 1993).

15. S. Johnson, V. Slaughter, and S. Carey, 'Whose Gaze Will Infants Follow? The Elicitation of Gaze-Following in Twelve-Month-Olds', *Developmental Science* 1 (1998): 233–8.

16. This example comes from A. N. Meltzoff and R. Brooks, 'Eyes Wide Shut: The Importance of Eyes in Infant Gaze Following and Understanding Other Minds', in *Gaze Following: Its Development and Significance*, edited by R. Flom, K. Lee, and D. Muir (Erlbaum, 2007).

17. B. M. Hood, J. D. Willen, and J. Driver, 'Adults' Eyes Trigger Shifts of Visual Attention in Human Infants', *Psychological Science* 9 (1998): 131–4.

18. A newborn's acuity is about one-twentieth of an adult's and would constitute a level of legal blindness.

19. T. Farroni, G. Csibra, F. Simion, and M. H. Johnson, 'Eye Contact Detection in Humans from Birth', *Proceedings of the National Academy of Sciences* 99 (2002): 9602–5.

20. S. M. J. Hains and D. W. Muir, 'Effects of Stimulus Contingency in Infant–Adult Interactions', *Infant Behaviour and Development* 19 (1996): 49–61.

21. M. Scaife and J. Bruner, 'The Capacity for Joint Visual Attention in the Infant', *Nature* 253 (1975): 265–6.

22. Danny Povinelli believes not. See Povinelli and Eddy, 'What Young Chimpanzees Know About Seeing' (see Note 2).

23. Barbara Smuts, 'What Are Friends For?', *Natural History* (American Museum of Natural History) (1987): 36–44.

24. V. Kuhlmeier, K. Wynn, and P. Bloom, 'Attribution of Dispositional States by Twelve-Month-Olds', *Psychological Science* 14 (2003): 402–8.

25. J. K. Hamlin, K. Wynn, and P. Bloom, 'Social Evaluation by Preverbal Infants', *Nature* 450 (2007): 557–9.

26. D. C. Dennett, 'Intentional Systems', *Journal of Philosophy* 68 (1971): 87–106.

27. D. C. Dennett, *Breaking the Spell: Religion as a Natural Phenomenon* (Penguin Allen Lane, 2006).

28. P. Bloom, *Descartes' Baby* (Basic Books, 2004).

29. B. Libet, 'Unconscious Cerebral Initiative and the Role of Conscious Will on Voluntary Action', *The Behavioural and Brain Sciences* 8 (1985): 529–66.

30. S. Pinker, *The Blank Slate: The Modern Denial of Human Nature* (Viking Adult, 2002), p. 43.

31. Descartes came to this conclusion because the pineal gland seemed to be one of the only structures in the brain that was not duplicated or organized in two halves. In fact, it is.

32. V. A. Ramachandran and S. Blakeslee, *Phantoms in the Brain: Probing the Mysteries of the Human Mind* (William Morrow, 1998).

33. D. M. Wegner, *The Illusion of Conscious Will* (MIT Press, 2002).

34. C. N. Johnson and H. M. Wellman, 'Children's Developing Conceptions of the Mind and Brain', *Child Development* 53 (1982): 222–34.

35. *Robocop*, directed by Paul Verhoeven (Orion Pictures, 1987).

36. *Blade Runner*, directed by Ridley Scott (Blade Runner Productions, 1982).

37. L. J. Rips, S. Blok, and G. Newman, 'Tracing the Identity of Objects', *Psychological Review* 113 (2006): 1–30.

38. V. Slaughter, 'Young Children's Understanding of Death', *Australian Psychologist* 40 (2005): 179–86.

39. J. M. Bering and D. F. Bjorkland, 'The Natural Emergence of Reasoning About the Afterlife as a Developmental Regularity', *Developmental Psychology* 40 (2004): 217–33.

40. J. M. Bering. 'Intuitive Conceptions of Dead Agents' Minds: The Natural Foundations of Afterlife Beliefs as Phenomenological Boundary', *Journal of Cognition and Culture* 2 (2002): 263–308.

41. V. Slaughter and M. Lyons, 'Learning About Life and Death in Early Childhood', *Cognitive Psychology* 46 (2002), 1–30.

42. J. M. Bering, C. Hernández-Blasi, and D. F. Bjorkland, 'The Development of 'Afterlife' Beliefs in Secularly and Religiously Schooled Children', *British Journal of Developmental Psychology* 23 (2005): 587–607.

43. J. M. Bering, 'The Folk Psychology of Souls', *The Behavioural and Brain Sciences* 29 (2006): 453–98.

CHAPTER SIX

1. Joseph Merrick is more commonly known as 'John Merrick' owing to a mistake resulting from the publication of the memoirs of his physician, Sir Frederick Treves.

2. Aloa, the Alligator Boy, was really William Smith who was born in Raleigh, North Carolina in 1908. He was the last of eight children. The seventh, his sister Virginia, was also born with the same skin condition. Aloa was seen by many doctors who attributed his condition to the fright his mother had received giving birth to his sister. He was most likely born with ichthyosis, a genetically inherited skin disorder.

3. The first case of a man with two penises (diphallia) was reported by Johannes Jacob Wecker in 1609. Diphallia has been estimated to occur in around 1 in 5.5 million male births in the United States; see K. K. Sharma, R. Jain, S. K. Jain, and A. Purohit, 'Concealed Diphallus: A Case Report and Review of the Literature', *Journal of the Indian Association of Pediatric Surgeons* 5 (2000): 18–21.

4. S. Carey, *Conceptual Change in Childhood* (Bradford Books of MIT Press, 1985).

5. S. A. Gelman, *The Essential Child: Origins of Essentialism in Everyday Thought* (Oxford University Press, 2003).

6. K. Inagaki and G. Hatano, 'Vitalistic Causality in Young Children's Naive Biology', *Trends in Cognitive Science* 8 (2004): 356–62.

7. Sir Hans Adolf Krebs won the Nobel Prize in 1953 for identifying the metabolic chemical reaction that produces energy in cells.

8. J. Lovelock, *Gaia: A New Look at Life on Earth* (Oxford University Press, 1979).

9. G. L. Murphey and D. L. Medin, 'The Role of Theories in Conceptual Coherence', *Psychological Review* 92 (1985): 289–316.

10. J. M. Mandler, *The Foundations of Mind* (Oxford University Press, 2004).

11. P. C. Quinn and P. D. Eimas, 'Perceptual Cues That Permit Categorical Differentiation of Animal Species by Infants', *Journal of Experimental Child Psychology* 63 (1996): 189–211.

12. S. Carey, 'Sources of Conceptual Change', in *Conceptual Development: Piaget's Legacy,* edited by E. K. Scholnick, K. Nelson, S. A. Gelman, and P. H. Miller (Erlbaum, 1999): 293–326.

13. S. Carey, 'Conceptual Differences Between Children and Adults', *Mind and Language* 3 (1988): 167–81.

14. For example, light in the infrared and ultraviolet ranges are beyond the limits of the human visual system. Likewise, humans can hear sounds only in the 20– to 20,000-hertz range.

15. *Invasion of the Body Snatchers*, directed by Don Siegel (Walter Wanger Productions, 1956).

16. D. L. Medin and A. Ortony, 'Psychological Essentialism', in *Similarity and Analogical Reasoning*, edited by S. Vosniadou and A. Ortony (Cambridge University Press, 1989).

17. The best and most accessible compilation is S. A. Gelman, *The Essential Child: Origins of Essentialism in Everyday Thought* (Oxford University Press, 2003).

18. S. A. Gelman and H. M. Wellman, 'Insides and Essences: Early Understandings of the Non-obvious', *Cognition* 38 (1991): 213–44.

19. L. A. Hirschfeld, 'Do Children Have a Theory of Race?', *Cognition* 54 (1995): 209–52.

20. S. A. Gelman and E. M. Markman, 'Categories and Induction in Young Children', *Cognition* 23 (1986): 183–209.

21. F. Keil, *Concepts, Kinds, and Cognitive Development* (Bradford Books, 1989).

22. J. H. Flavell, E. R. Flavell, and F. L. Green, 'Development of the Appearance-Reality Distinction', *Cognitive Psychology* 15 (1983): 95–120.

23. G. E. Newman and F. C. Keil, 'Where's the Essence?': Developmental Shifts in Children's Beliefs About Internal Features' (*Child Development*, in press).

24. In fact, the idea is not to eat the potato at all. Rather, Professor Tony Trewavas of the Edinburgh Institute of Molecular Plant Sciences developed the genetically modified potato as a marker plant that could be used to monitor the crop as a whole. Simply by sowing a couple of plants in the field, the farmer would be able to regulate watering and improve yields of the normal potatoes.

25. H. Bagis, D. Aktoprakligil, H. O. Mercan, N. Yurusev, G. Turget, S. Sekman, S. Arat, and S. Cetin, 'Stable Transmission and Transcription of Newfoundland Ocean Pout Type III Fish Antifreeze Protein (AFP) Gene in Transgenic Mice and Hypothermic Storage of Transgenic Ovary and Testis', *Molecular Reproduction and Development* 73 (2006): 1404–11.

26. *The Fly*, directed by David Cronenberg (Brooksfilms, 1986).

27. P. Savolainen, Y. Zhang, J. Luo, J. Lundeberg, and T. Leitner, 'Genetic

Evidence for an East Asian Origin of Domestic Dogs', *Science* 298 (2002): 1610–13.

28. Stem cells come in two forms, embryonic and adult. Adult stem cell therapies are relatively uncontroversial and have been used for many years in the treatment of leukemia. In contrast, human embryonic stem cells are potentially capable of regenerative treatment for a wider variety of damaged and diseased cell conditions, but, because they involve the destruction of embryos, research and practice are still highly controversial and banned in many countries.

29. The original study published by Joseph Vacanti and his colleagues in *Plastic and Reconstructive Surgery* in 1997 caused a storm of public outrage and confused controversy. In 1999 the anti-genetic engineering group Turning Point Project took out an ad in the *New York Times* showing a picture of the mouse with the misleading caption, 'This is an actual photo of a genetically engineered mouse with a human ear on its back.' The mouse was not genetically engineered, nor were there any human cells implanted in it. In fact, the bioframe was made from cow cartilage.

30. The belief in vital life forces and energies is found in most Eastern philosophies. For a discussion of Western notions of vitalism, see E. Mayr, *The Growth of Biological Thought* (Harvard University Press, 1982).

31. M. Roach, *Six Feet Over: Adventures in the After Life* (Cannongate, 2007).

32. D. Macdougall, 'Hypothesis Concerning Soul Substance Together with Experimental Evidence of the Existence of Such Substance', *American Medicine* 4 (1907): 240–43.

33. K. Inagaki and G. Hatano, *Young Children's Naive Thinking About the Biological World* (Psychology Press, 2002).

34. V. Slaughter and M. Lyons, 'Learning About Life and Death in Early Childhood', *Cognitive Psychology* 43 (2003): 1–30.

35. 'Quintessence' as a term survives today in modern theoretical physics as the name for the hypothetical dark energy that is believed to account for the energy necessary to explain the continuing expansion of the known universe.

36. For an accessible introduction to the Great Chain of Being and the emergence of the scientific method out of the age of alchemy, I recommend J. Henry, *Knowledge Is Power: How Magic, the Government, and an Apocalyptic Vision Inspired Francis Bacon to Create Modern Science* (Icon Books, 2002).

37. B. Woolley, *The Herbalist: Nicholas Culpeper and the Fight for Medical Freedom* (HarperCollins, 2004).
38. The coco de mer is a rare protected member of the palm species that grows only in the Seychelles Islands. It used to be thought to resemble a woman's buttocks, which is reflected in one of its old botanical names, *Lodoicea callipyge*, in which *callipyge* is from the Greek for 'beautiful rump'.
39. See Andrew Harding, 'Beijing's Penis Emporium', *BBC News*, 23 September, 2006, available at: http://news.bbc.co.uk/1/hi/programmes/from_our_own_correspondent/5371500.stm.
40. Ben Goldacre, the scourge of quacks and charlatans has covered homeopathy in his eminently readable, *Bad Science* (Fourth Estate, 2008).
41. Tony Tysome, 'Rise in Applications for "Soft" Subjects Panned as Traditional Courses Lose Out', *Times Higher Education Supplement*, 27 July, 2007, available at http://www.timeshighereducation.co.uk/story.asp?storyCode=209755&sectioncode=26
42. Meirion Jones, 'Malaria Advice Risks Lives', *BBC News*, 13 July, 2006, available at: http://news.bbc.co.uk/1/hi/programmes/newsnight/5178122.stm.
43. M. Sans-Corrales, E. Pujol-Ribera, J. Gené-Badia, M. I. PasarínRua, B. Iglesias-Pérez, and J. Casajuana-Brunet, 'Family Medicine Attributes Related to Satisfaction, Health, and Costs', *Family Practice* 23 (2006): 308–16.
44. P. Rozin, L. Millman, and C. Nemeroff, 'Operation of the Laws of Sympathetic Magic in Disgust and Other Domains', *Journal of Personality and Social Psychology* 50 (1986): 703–12.
45. B. Wicker, C. Keysers, J. Plailly, J. P. Royet, V. Gallese, and G. Rizzolatti, 'Both of Us Disgusted in My Insula: The Common Neural Basis of Seeing and Feeling Disgusted', *Neuron* 40 (2003): 655–64.
46. C. Nemeroff and P. Rozin, 'The Contagion Concept in Adult Thinking in the United States: Transmission of Germs and of Interpersonal Influence', *Ethos* 22 (1994): 158–86.
47. M. Mauss, *A General Theory of Magic*, translated by R. Brain (1902; reprint, W. W. Norton, 1972).
48. P. Rozin and A. Fallon, 'The Acquisition of Likes and Dislikes for Foods', in *What Is America Eating? Proceedings of a Symposium* (National Academies Press, 1986), available at: http://www.nap.edu/openbook/0309036356/html/58.html.

49. L. R. Kass, 'The Wisdom of Repugnance', *The New Republic* (2 June, 1997): 17–26.

50. J. Haidt, S. H. Koller, and M. G. Dias, 'Affect, Culture, and Morality, or Is It Wrong to Eat Your Dog?' *Journal of Personality and Social Psychology* 65 (1993): 613–28.

51. J. Haidt, 'The Emotional Dog and Its Rational Tail: A Social Intuitionist Approach to Moral Judgement', *Psychological Review* 108 (2001): 814–34.

CHAPTER SEVEN

1. S. Schachter and J. E. Singer, 'Cognitive, Social, and Physiological Determinants of Emotional States', *Psychological Review* 69 (1962): 379–99.

2. D. G. Dutton and A. P. Aron, 'Some Evidence for Heightened Sexual Attraction Under Conditions of High Anxiety', *Journal of Personality and Social Psychology* 30 (1974): 510–17.

3. 'Untouchables' refers to the lowest castes in several different societies, including 'Baekjeong' (Korea), 'Burakumin' (Japan), 'Khadem' (Yemen), and similar castes in many African countries. Although Western countries may have abandoned official social segregation, seating arrangements on some forms of public transport and in arenas of public entertainment retain the legacy of maintaining a physical distance between the upper and lower classes.

4. D. Rothbart and T. Barlett, 'Rwandan Radio Broadcasts and Hutu/Tutsi Positioning', in *Conflict and Positioning Theory*, edited by F. M. Moghaddam and R. Harré (Springer, 2007).

5. R. T. McNally, *Dracula Was a Woman: In Search of the Blood Countess of Transylvania* (McGraw-Hill, 1987). For a rebuttal of this theory, see E. Miller, *Dracula: Sense and Nonsense* (Parkstone Press, 2000).

6. T. Thorne, *Countess Dracula: The Life and Times of the Blood Countess, Elisabeth Báthory* (Bloomsbury, 1997).

7. Peta Bee, 'Naturally Dangerous?', *The Times*, 16 July, 2007, available at: http://www.timesonline.co.uk/tol/life_and_style/health/features/article2073171.ece

8. 'Ask Hugh', available at River Cottage website: http://www .rivercottage.net/askhugh.

9. 'She's Her Own Twin', *ABCNews*, 15 August, 2006, available at: http://abcnews.go.com/Primetime/story?id=2315693.

10. C. Ainsworth, 'The Stranger Within', *New Scientist* 180 (2003): 34.

11. N. Yu, M. S. Kruskall, J. J. Yunis, J. H. M. Knoll, L. Uhl, S. Alosco, M. Ohashi, O. Clavijo, Z. Husain, and E. J. Yunis, 'Disputed Maternity Leading to Identification of Tetragametic Chimerism', *New England Journal of Medicine* 346 (2002): 1545–52.

12. W. Arens, *The Man-Eating Myth: Anthropology and Anthropophagy* (Oxford University Press, 1979); see also G. Obeyesekere, *Cannibal Talk: The Man-Eating Myth and Human Sacrifice in the South Seas* (University of California Press, 2005). For a rebuttal, see T. White, *Prehistoric Cannibalism at Mancos 5Mtumr-2346* (Princeton University Press, 1992).

13. Carlton Gajdusek received the 1976 Nobel Prize for Medicine for his discovery of the prion disease pattern Kuru in the Fore tribe.

14. R. L. Klitzman, M. Alpers, and D. C. Gajdusek, 'The Natural Incubation Period of Kuru and the Episodes of Transmission in Three Clusters of Patients', *Neuroepidemiology* 3 (1984): 3–20.

15. R. A. Marlar, B. L. Leonard, B. R. Billman, P. M. Lambert, and J. E. Marlar, 'Biochemical Evidence of Cannibalism at a Prehistoric Puebloan Site in Southwestern Colorado', *Nature* 407 (2000): 74–8.

16. Different reasons are given for ceremonial cannibalism and the practices associated with it. Most of these interpretations are based on interviews with surviving tribe members, since cannibalistic practices have generally been outlawed since the 1960s. For an account of the Wari of South America and funerary cannibalism, read Beth Corkin, *Consuming Grief: Compassionate Cannibalism in Amazonian Society* (University of Texas, 2001). The practices of the Melanesian Kukukukus are documented in Jens Bjerre, *The Last Cannibals* (Michael Joseph, 1956).

17. Luke Harding, 'Victim of Cannibal Agreed to Be Eaten', the *Guardian*, 4 December, 2003, available at: http://www.guardian.co.uk/germany/article/0,2763,1099477,00.html. Transcripts of the trial are available in G. Stampf, 'Interview mit einem Kannibalen', *Gebundene Ausgabe* (2007).

18. 'Interview with a Cannibal', RDF Media/Stampfwerk coproduction for Five (2007).

19. I first came across the Gammons' case in 'Help! I'm Turning into My Wife', *Daily Mail*, 9 November, 2006, available at: http://www.dailymail.co.uk/pages/live/femail/article.html?in_article_id=415584&in_page_id=1879. I have since spoken by phone with Ian, who verifies the account reported in the paper.

20. Y. Inspector, I. Kutz, and D. David, 'Another Person's Heart: Magical and

Rational Thinking in the Psychological Adaptation to Heart Transplantation', *Israel Journal of Psychiatry and Related Sciences* 41 (2004): 161–73.

21. C. Sylvia and W. Novak, *A Change of Heart: A Memoir* (Grand Central Publishing, 1998).

22. J. V. McConnell, 'Memory Transfer Through Cannibalism in Planarians', *Journal of Neurophysiology* 3 (1962): 42–8.

23. G. Ungar, L. Galvan, and R. H. Clark, 'Chemical Transfer of Learned Fear', *Nature* 217 (1968): 1259–61.

24. B. Frank, D. G. Stein, and J. Rosen, 'Interanimal Memory Transfer: Results from Brain and Liver Homogenates', *Science* 169 (1970): 399–402.

25. P. P. Pearsall, *The Heart's Code* (Broadway, 1999).

26. C. Dyer, 'English Teenager Given Heart Transplant Against Her Will', *British Medical Journal* 319, no. 7204 (1999): 209.

27. M. A. Sanner, 'People's Feelings and Ideas About Receiving Transplants of Different Origins: Questions of Life and Death, Identity, and Nature's Border', *Clinical Transplantation* 15 (2001): 19–27; M. A. Sanner, 'Exchanging Spare-Parts or Becoming a New Person? People's Attitudes Toward Receiving and Donating Organs', *Social Science Medicine* 52 (2001): 1491–99; M. A. Sanner, 'Giving and Taking – to Whom and from Whom? People's Attitudes Toward Transplantation of Organs and Tissue from Different Sources', *Clinical Transplantation* 12 (1998): 515–22; M. A. Sanner, 'Living with a Stranger's Organ: Views of the Public and Transplant Recipients', *Annals of Transplantation* 10 (2005): 9–12.

28. B. M. Hood, K. Donnelly, and A. Byers, 'Moral Contagion and the Horns Effect: Attitudes Towards Potential Organ Transplant', *Journal of Culture and Cognition* (in press).

29. 'New Rules on Organ Donation', *BBC News*, 22 February, 2000, available at: http://news.bbc.co.uk/1/hi/health/651270.stm.

30. R. M. Veatch, *Transplantation Ethics* (Georgetown University Press, 2000).

31. M. Sanner, 'Transplant Recipients' Conceptions of Three Key Phenomena in Transplantation: The Organ Donation, the Organ Donor, and the Organ Transplant', *Clinical Transplantation* 17 (2003): 391–400.

32. *Barbarella*, directed by Roger Vadim (Dino de Laurentiis Cinematografica, 1968).

33. As reported 4 June 2008. At http://edition.cnn.com/2008/WORLD/europe/06/04/cathedral.sex/index.html

34. M. Earl-Taylor, 'HIV/AIDS, the Stats, the Virgin Cure, and Infant Rape', *Science in Africa* (April 2002), available at:

http://www.scienceinafrica.co.za/2002/april/virgin.htm.

35. G. J. Pitcher and D. M. Bowley, 'Infant Rape in South Africa', *The Lancet* 359 (2002): 274–5.

36. C. H. Legare and S. A. Gelman, 'Bewitchment, Biology, or Both: The Coexistence of Natural and Supernatural Explanatory Frameworks Across Development', *Cognitive Science* (in press).

37. C. MacKay, *Extraordinary Popular Delusions and the Madness of Crowds* (1841; reprint, Wordsworth, 1995).

38. L. R. Alton, *The Bizarre Careers of John R. Brinkley* (University Press of Kentucky, 2002).

39. http://www.time.com/time/magazine/article/0,9171,727231,00.html

40. R. V. Short, 'Did Parisians Catch HIV from Monkey Glands?', letter to *Nature* 398 (1999): 659.

41. This account of the execution comes from the memoirs of a clergyman, Philip Henry (1631–96). However, no other account mentions this crowd response.

42. See the Auld Sod Export Company Ltd. webside at: http://89.234.45.183/index.htm.

43. See 'About Claridge's' at: http://www.claridges.co.uk/about_claridges/history.

CHAPTER EIGHT

1. Pamela Wiggins, 'Top Eight Celebrity Collectibles and Who Collects Them', available at: About.com, http://antiques.about.com/od/showntell/tp/aa012807.htm.

2. B. N. Frazier, S. A. Gelman, A. Wilson, and B. Hood, 'Picasso Paintings, Moon Rocks, and Handwritten Beatle Lyrics: Adults' Evaluations of Authentic Objects', unpublished paper.

3. Mariusz Lodkowski, 'Battle over a Suitcase from Auschwitz', *Sunday Times*, 13 August, 2006, available at: http://www.timesonline .co.uk/tol/news/world/article607646.ece.

4. The Native American Graves Protection and Repatriation Act of 1990 (NAGPRA) protects Native American burial sites and also enables Native Americans to reposses ancestral remains held by museums and other scientific institutions.

5. The Lascaux caves were discovered in 1940, but by 1955 the carbon

dioxide of visitors had visibly destroyed the paintings, and so it was closed to the public in 1963. In 1983 Lascaux II, a reconstruction, was opened two hundred metres from the actual caves.

6. F. Wynne, *I Was Vermeer: The Forger Who Swindled the Nazis* (Bloomsbury, 2006).

7. Chris Gray, 'Bloody Hell: A Headache for Saatchi as Prize Artwork Defrosts', *The Independent*, 4 July, 2002, available at: http://news.independent.co.uk/uk/this_britain/article182737.ece.

8. Jenny Booth and Nico Hines, 'We Can Save the Cutty Sark After Blaze, Say Ship's Owners', *The Times*, 21 May, 2007, available at: http://www.timesonline.co.uk/tol/news/uk/article1817806.ece.

9. D. G. Hall, 'Continuity and Persistence of Objects', *Cognitive Psychology* 37 (1998): 28–59.

10. Nicholas Wade, 'Your Body Is Younger Than You Think', *New York Times*, 2 August, 2005, available at: http//www.nytimes.com/2005/08/02/science/02cell.html?pagewanted=1&_r=1.

11. 'Big Girls Don't Cry', lyrics and music by Fergie and Toby Gad (A&M Records, 2007).

12. M. Hobra, 'Prevalence of Transitional Objects in Young Children in Tokyo and New York', *Infant Mental Health Journal*, 24, (2003): 174-91.

13. Travelodge press release, 13 March, 2007, available at: http://www.travelodge.co.uk/press/article.php?id=222.

14. L. M. Krauss, *The Physics of 'Star Trek'* (HarperCollins, 1996).

15. B. M. Hood and P. Bloom, 'Children Prefer Certain Individuals over Perfect Duplicates', *Cognition* (2008): 455–62.

16. P. Bloom and S. Gelman, "Psychological essentialism in selecting the 14th Dalai Lama," *Trends in Cognitive Sciences*, 12, (2008): 243. Article based on K.S. Wangdu, 'Report on the Discovery, Recognition, and Enthronement of the Fourteenth Dalai Lama.' New Delhi: Government of India Press. Reprinted in 'Discovery, Recognition, and Enthronement of the Fourteenth Dalai Lama: A Collection of Accounts', edited by Library of Tibetan Work & Archives, New Delhi: Indraprastha Press (1941).

17. C. N. Johnson and M. G. Jacobs, 'Enchanted Objects: How Positive Connections Transform Thinking About the Nature of Things', poster and presentation at the symposium 'Children's Thinking About Alternative Realities' (C. Johnson, chair), biennial meeting of the

Society for Research in Child Development, Minneapolis, Minn., 19 April, 2001.

18. *The Prestige*, directed by Christopher Nolan (Newmarket Productions, 2006).

19. Nikola Tesla (1856–1943) was a Serbian responsible for inventing alternating electrical current.

20. B. M. Hood and P. Bloom, 'Do Children Think That Duplicating the Body Also Duplicates the Mind?', unpublished paper.

21. J. Capgras and J. Reboul-Lachaux, 'L'Illusion des soles dans un delire systematise chronique', *Bulletin de Society Clinique de Medicine Mentale* 11 (1923): 6–16.

22. G. Blount, 'Dangerousness of Patients with Capgras Syndrome', *Nebraska Medical Journal* 71 (1986): 207.

23. V. A. Ramachandran and S. Blakeslee, *Phantoms in the Brain: Probing the Mysteries of the Human Mind* (William Morrow, 1998).

24. A. Ghaffari-Nejad and K. Toofani, 'A Report of Capgras Syndrome with Belief in Replacement of Inanimate Objects in a Patient Who Suffered from Grandmal Epilepsy', *Archives of Iranian Medicine* 8 (2005): 141–3.

25. T. Feinberg, *Altered Egos: How the Brain Creates the Self* (Oxford University Press, 2000).

26. R. T. Abed and W. D. Fewtrell, 'Delusional Misidentification of Familiar Inanimate Objects: A Rare Variant of Capgras Syndrome', *British Journal of Psychiatry* 157 (1990): 915–17.

27. H. D. Ellis and M. B. Lewis, 'Capgras Delusion: A Window on Face Recognition', *Trends in Cognitive Science* 5 (2001): 149–56.

CHAPTER NINE

1. This example is taken from R. Sheldrake, *The Sense of Being Stared At: And Other Aspects of the Extended Mind* (Arrow Books, 2004).

2. C. G. Gross, 'The Fire That Comes from the Eye', *The Neuroscientist* 5 (1999): 58–64.

3. This discovery was first articulated experimentally by the Arabic scholar Alhazen, who invented the pinhole camera and explained why the image was inverted because of the optics of light entering the eye.

4. Red eye is due to the reflection of blood vessels that cover the surface of

the back of the eye. The light-sensitive surface of the back of the eye, known as the retina, is actually organized back to front, with light having to pass through the blood supply before reaching the light receptors.

5. G. A. Winer, J. E. Cottrell, V. Gregg, J. S. Fournier, and L. A. Bica, 'Fundamentally Misunderstanding Visual Perception: Adults' Belief in Visual Emissions', *American Psychologist* 57 (2002): 417–24.

6. S. Freud, *Group Psychology and the Analysis of the Ego* (1921; reprint, W. W. Norton, 1975).

7. T. Depoorter, 'Madame Lamort and the Ultimate Medusa Experience', *Image and Narrative: Online Magazine of the Visual Narrative*, issue 5: 'The Uncanny' (January 2003), available at: http://www.imageandnarrative.be/uncanny/treesdepoorter.htm. 'It is noteworthy that in different versions of the ancient myth of Medusa, it is sometimes the sight of her – her "being gazed at" by a spectator, while at other times it is the gaze of Medusa herself, her "looking at" a spectator that petrifies.'

8. In February 2008, the appropriately named Third Penal Division of the Rome court ruled it a criminal offense for Italian men to touch their genitals in public. The ban did not just apply to brazen crotch-scratching, but also to the superstitious practice to ward off evil. See John Hooper, 'Touch Your Privates in Private, Court Tells Italian Men', the *Guardian*, 28 February, 2008, available at: http://www.guardian.co.uk/world/2008/feb/27/italy1.

9. B. Castiglione, *The Book of the Courtier*, translated by C. Singleton (1528; reprint, Anchor Books, 1959).

10. C. Russ, 'An Instrument Which Is Set in Motion by Vision or by Proximity of the Human Body', *The Lancet* 201 (1921): 222–34.

11. J. Ronson, *The Men Who Stare at Goats* (Picador, 2004).

12. E. B. Titchener, 'The Feeling of Being Stared At', *Science* (new series) 308 (1898): 23.

13. This comes from an unpublished study I conducted with 219 first-year students who had taken courses in perception and vision at the University of Bristol. In addition to filling out a standard questionnaire that measured paranormal beliefs (T. M. Randall, 'Paranormal Short Inventory', *Perceptual and Motor Skills* 84 [1997]: 1265–6), they were asked to rate the statement, 'People can tell when they are being watched even though they cannot see who is watching them' on a scale of 1 (strong disagreement) to 6 (strong agreement). Only 4 per cent rated 'strong

disagreement', and 9 per cent rated 'disagreement.' The remainder of the students agreed with this statement to some extent, even though, as a group, they scored lower than other large samples for paranormal beliefs.

14. This example is taken from R. Sheldrake, *The Sense of Being Stared At: And Other Aspects of the Extended Mind* p. 178 (see Note 1).

15. P. Brugger and K. I. Taylor, 'ESP: Extrasensory Perception or Effect of Subjective Probability?', *Journal of Consciousness Studies* 10 (2003): 221–46.

16. J. Colwell, S. Schröder, and D. Sladen, 'The Ability to Detect Unseen Staring: A Literature Review and Empirical Tests', *British Journal of Psychology* 91 (2000): 71–85.

17. The phrase is attributed to Carl Sagan, though he was paraphrasing a statement originally made by David Hume. For discussion, see M. Pigliucci, 'Do Extraordinary Claims Really Require Extraordinary Evidence?', *The Sceptical Inquirer* (March–April 2005).

18. J. H. Flavell, 'Development of Knowledge About Vision', in *Thinking and Seeing: Visual Metacognition in Adults and Children*, edited by D. T. Levin (MIT Press, 2004).

19. A. A. di Sessa, 'Towards an Epistemology of Physics', *Cognition and Instruction* 10 (1993): 105–225.

20. J. E. Cottrell and G. A. Winer, 'Development in the Understanding of Perception: The Decline of Extramission Perception Beliefs', *Developmental Psychology* 30 (1994): 218–28.

21. S. Einav and B. M. Hood, 'Children's Use of Temporal Dimension of Gaze to Infer Preference', *Developmental Psychology* 42 (2006): 142–52.

22. S. Baron-Cohen, *Mindblindness* (MIT Press, 1995).

23. R. B. Adams, H. L. Gordon, A. A. Baird, N. Ambady, and R. E. Kleck, 'Effects of Gaze on Amygdala Sensitivity to Anger and Fear Faces', *Science* 300 (2003): 1536.

24. K. Nicholas and B. Champness, 'Eye Gaze and the GSR', *Journal of Experimental Social Psychology* 7 (1971): 623–6.

25. M. Argle and M. Cook, *Gaze and Mutual Gaze* (Cambridge University Press, 1976).

26. This explanation was first suggested by Titchener, 'The Feeling of Being Stared At.'

27. R. Sheldrake, *A New Science of Life: The Hypothesis of Morphic Resonance* (Park Street Press, 1981).

28. According to Sheldrake's theory, any system, including minds, can assume a particular shape or configuration. 'Morphic' means shape. A change in

the shape of any one system affects the collective shape of all related systems. This is the resonance part of the theory. Subsequent systems resonate with other systems, enabling information to travel across space and time. The effect is stronger the more systems are involved and the more similar the future system is to the systems that generated the field. In 1989 the experimental psychologist Zoltan Dienes undertook research to investigate morphic resonance by testing it with remote viewing of repetition priming. In repetition priming, people respond more quickly and accurately with repeated presentation. He wanted to know if people trained on word recognition tasks influenced a different group of people through the effects of thought transference. Initially, he found a significant effect. Unfortunately, when he ran the study another two times, there was no effect. Dienes explains in mathematical detail why some theories should be tested and why others should not. More importantly, he explains the importance and difficulty of establishing truth in his new book, *Understanding Psychology as a Science: An Introduction to Scientific and Statistical Inference* (Palgrave Macmillan, 2008).

29. G. Orwell, *Nineteen Eighty-Four: A Novel* (Secker & Warburg, 1949).

30. M. Milinski, D. Semmann, and H. J. Krambeck, 'Reputation Helps Solve "Tragedy of the Commons"', *Nature* 415 (2002): 424–6.

31. M. Bateson, D. Nettle, and G. Roberts, 'Cues of Being Watched Enhance Cooperation in a Real World Setting', *Biology Letters* 2 (2006): 412–14.

32. J. M. Bering, K. A. McLeod, and T. K. Shackelford, 'Reasoning About Dead Agents Reveals Possible Adaptive Trends', *Human Nature* 16 (2005): 360–81.

33. J. M. Bering, 'The Folk Psychology of Souls', *The Behavioural and Brain Sciences* 29 (2006): 453–98.

34. S. Guthrie, *Faces in the Clouds: A New Theory of Religion* (Oxford University Press, 1993).

35. K. Conrad, *Die beginnende Schizophrenie: Versuch einer Gestaltanalyse des Wahns* (Thieme, 1958).

36. T. C. Manschreck, B. A. Maher, J. J. Milavetz, D. Ames, C. C. Weisstein, and M. L. Schneyer, 'Semantic Priming in Thought Disordered Schizophrenic Patients', *Schizophrenia Research* 1 (1988): 61–6.

37. M. Eckblad and L. J. Chapman, 'Magical Ideation as an Indicator of Schizotypy', *Journal of Consulting and Clinical Psychology* 51 (1983): 215–25.

38. B. E. Brundage, 'First-Person Account: What I Wanted to Know but Was Afraid to Ask', *Schizophrenia Bulletin* 9 (1983): 583–5, 584.

39. R. Montague, *Why Choose This Book? How We Make Decisions* (Dutton, 2007).

40. G. Fénelon, F. Mahieux, R. Huon, and M. Ziégler, 'Hallucinations in Parkinson's Disease', *Brain* 123 (2000): 733–45.

41. R. King, J. D. Barchas, and B. A. Huberman, 'Chaotic Behaviour in Dopamine Neurodynamics', *Proceedings of the National Academy of Sciences* 81 (1984): 1244–7; see also A. Shaner, 'Delusions, Superstitious Conditioning, and Chaotic Dopamine Neurodynamics', *Medical Hypothesis* 52 (1999): 119–23.

42. P. Brugger, 'From Haunted Brain to Haunted Science: A Cognitive Neuroscience View of Paranormal and Pseudoscientific Thought', in *Hauntings and Poltergeists: Multidisciplinary Perspectives*, edited by J. Houran and R. Lange (McFarland & Co., 2001).

43. S. Blackmore and T. Troscianko, 'Belief in the Paranormal: Probability Judgements, Illusory Control, and the Chance Baseline Shift', *British Journal of Psychology* 76 (1985): 459–68.

44. P. Krummenacher, P. Brugger, M. Fathi, and C. Mohr, 'Dopamine, Paranormal Ideation, and the Detection of Meaningful Stimuli', Paper presented at the Zentrum fur Neurowissenschaften, Zurich (2002).

45. P. Goldman-Rakic, 'Working Memory and the Mind', in *Mind and Brain: Readings from Scientific American* (W. H. Freeman & Co., 1993).

46. A. Baddeley, *Working Memory* (Oxford University Press, 1986).

47. J. Emick and M. Welsh, 'Association Between Formal Operational Thought and Executive Function as Measured by the Tower of Hanoi-Revised', *Learning and Individual Differences* 15 (2005): 177–88.

48. J. R. Stroop, 'Studies of Interference in Serial Verbal Reactions', *Journal of Experimental Psychology* 18 (1935): 643–62. This test is usually done with words printed in different colours, but my publishers warned me that the cost of printing just a few words in colour could not be justified. Luckily for me, Steve Pinker presumably encountered the same issue in his latest book, *The Stuff of Thought* (2007, p. 332), where he also describes the Stroop effect. I have used his solution in overcoming the use of coloured ink to achieve the same demonstration.

49. A. Diamond, *The Development and Neural Bases of Higher Cognitive Functions* (New York Academy of Sciences, 1990).

50. K. Dunbar, J. Fugelsang, and C. Stein, 'Do Naive Theories Ever Go Away?' in *Thinking with Data: Thirty-third Carnegie Symposium on Cognition*, edited by M. Lovett and P. Shah (Erlbaum, in press).

51. In fact, it is the right dorsal lateral prefrontal cortex which is active when adults are actively suppressing the urge to respond with the intuitive response. Wim De Neys, Oshin Vartanian, Vinod Goel (2008) 'Smarter Than We Think: When Our Brains Detect That We Are Biased', *Psychological Science* 19 (5), 483–9

52. T. Lombrozo, D. Kelemen, and D. Zaitchik, 'Teleological Explanation in Alzheimer's Disease Patients', *Psychological Science* 18 (2007): 999–1006.

53. D. Zaitchik and G. Solomon, 'Animist Thinking in the Elderly and in Patients with Alzheimer's Disease', *Cognitive Neuropsychology* (in press).

54. J. S. B. T. Evans, 'In Two Minds: Dual Process Accounts of Reasoning', *Trends in Cognitive Science* 7 (2003): 454–9.

55. S. Epstein, R. Pacini, V. Denes-Raj, and H. Heier, 'Difference in Intuitive–Experiential and Analytical–Rational Thinking Styles', *Journal of Personality and Social Psychology* 71 (1996): 390–405.

56. L. A. King, C. M. Burton, J. A. Hicks, and S. M. Drigotas, 'Ghosts, UFOs, and Magic: Positive Affect and the Experiential System', *Journal of Personality and Social Psychology* 92 (2007): 905–19.

57. M. Lindeman and K. Aarnio, 'Superstitious, Magical, and Paranormal Beliefs: An Integrative Model', *Journal of Research in Personality* 41 (2007): 731–44; M. Lindeman and M. Saher, 'Vitalism, Purpose, and Superstition', *British Journal of Psychology* 98, no. 1 (2007): 33–44; M. Lindeman and K. Aarnio, 'Paranormal Beliefs: Their Dimensionality and Psychological Correlates', *European Journal of Personality* 20 (2006): 585–602.

CHAPTER TEN

1. P. E. Tetlock, 'Thinking the Unthinkable: Sacred Values and Taboo Cognitions', *Trends in Cognitive Science* 7 (2003): 320–24.

2. W. Styron, *Sophie's Choice* (Random House, 1979).

INDEX

Illustrations or photographs are indicated by bold numbers

babies (*cont.*)
  90, 101–2; knowledge (of
  world) 105–6; 'Little Albert'
  experiments 67, 68–9; magic
  tricks 102–5; mind-reading
  122; and missing square
  pattern 24–5; and moral
  behaviour 132; movement
  128–9; reality, sense of 101–2;
  reasoning 63–4, 106; 'shapka'
  experiment 129; stepping
  reflex 88–9; startle reflex 90;
  sucking reflex 88; unborn 91
  See also children; newborns
babyness 123
'Babies are Smarter Than You
  Think' (*Life*) 94
Barbarella 197–8
Baron-Cohen, Simon 122
Barrett, Justin 62
Bartlett, Sir Frederick 100
Báthory, Countess Erzsebet
  184–5
Beckham, David 30
behaviour: conditioning
  mechanism 96–8; irrational
  98; moral 131–3; obsessive
  30–1, 67; staring 236–8,
  240–2 *See also* rituals
behaviourism 97–100
beliefs 19–23, 261–3; afterlife
  145–7, 247; 'apparent mental
  causation' 32–3; children's
  fantasy figures 20, 63; culture,
  spread by 20–1, 57–61; female

bank worker example 65–6;
  intuitive 48; phobias 66–9;
  sacred 11–14, 84–5, 266–7;
  taboos 178; unseen gaze
  236–8, 240–2; vision, regarding
  237, 262 *See also* religion
Bering, Jesse 247
bestiality 43–4
biology: children's understanding
  of 155–7; DNA 152, 171,
  187–9, 192–7; essence of life
  158–9; freaks 149–53; genetic
  engineering 161–4; mosaicism
  148, 189 *See also* animals
Blackmore, Susan 251
Blade Runner 144
Blanc, Lori 84
blankets, security 220–5
Bloom, Paul 135, 140, 224
Bouchard, Thomas 82
Boyer, Pascal 69–70
brain: Alzheimer's disease
  258–9; amygdala 243; DLPC
  254–8; children's understanding of 142–3;
  dopamine system 251–3;
  environment and 80; and
  facial recognition 124; frontal
  lobe 41–2, 90; fusiform gyrus
  80, 124; insula 176; limbic
  system 41; and mind 139–41,
  147–8; natural reflexes of the
  88–90; neurotransmitters 251;
  old age and 257–8; and
  pattern-seeking 24–7; pineal